Metal Magnetic Memory Technique and Its Applications in Remanufacturing

C000156738

Haihong Huang · Zhengchun Qian · Zhifeng Liu

Metal Magnetic Memory Technique and Its Applications in Remanufacturing

Haihong Huang
School of Mechanical Engineering
Hefei University of Technology
Hefei, Anhui, China

Zhengchun Qian
School of Mechanical Engineering
Nanjing Institute of Technology
Nanjing, Jiangsu, China

Zhifeng Liu
School of Mechanical Engineering
Hefei University of Technology
Hefei, Anhui, China

ISBN 978-981-16-1592-4 ISBN 978-981-16-1590-0 (eBook)
https://doi.org/10.1007/978-981-16-1590-0

Jointly published with Science Press
The print edition is not for sale in China (Mainland). Customers from China (Mainland) please order the print book from: Science Press.

This Springer imprint is published by the registered company Springer Nature Singapore Pte Ltd.
The registered company address is: 152 Beach Road, #21-01/04 Gateway East, Singapore 189721, Singapore

Preface

This book introduces the Metal Magnetic Memory (MMM) technique, one of the developing nondestructive testing methods, and its applications in remanufacturing engineering. MMM technique can effectively evaluate the stress concentration, plastic deformation, crack length and fatigue life of ferromagnetic materials. The advantages of this technique, including high-speed, low-cost and without preprocessing, determine that it is suitable to evaluate the early damage degree of remanufacturing cores, as well as the repair quality of remanufactured components.

The various MMM signal characteristics are extracted in this book to reflect the damage degree of remanufacturing cores, coatings and interfaces. In order to improve the accuracy of evaluation, the MMM signals induced by various damages, as well as applied stress and frictional wear, are analyzed. Besides, the effects of environmental temperature and excitation field on MMM signals are discussed and the stress concentration is proved to be the underlying reason for spontaneous magnetization. All the theoretical models, analysis methods and testing results are discussed and summarized to provide guidance to control the quality of remanufactured parts and products. The aim of this book is to help the academics for developing the novel MMM technique and its applications in remanufacturing, and also to serve the students, teachers, researchers, engineers and other people, who are interested in the magnetic nondestructive testing and the quality control of remanufacturing process.

In the past few years, the authors have conducted in-depth studies on the MMM nondestructive testing and devoted to promoting the application of MMM technique to remanufacturing engineering. The main research achievements are summarized in this book and the main contents are as follows:

- Reviews the motivations, the fundamentals and the recent research advances of the MMM technique, and explains the necessity of MMM technique for the quality control and evaluation in remanufacturing.
- Evaluates the damage degrees of remanufacturing cores induced from applied stress and frictional wear by the MMM signals, and discusses the effects of stress concentration, excitation magnetic field and high-temperature environment on the MMM testing results.

- Evaluates the heat residual stress and fatigue life of remanufactured coatings by MMM signals, and establishes a new magnetomechanical coupling model for the coating/substrate interface damage evaluation.
- For the typical components, including the retired drive axle housing and hydraulic cylinder and the remanufactured engine crankshaft, the MMM technique is used to evaluate the damage degree of remanufacturing cores and the repair quality of remanufactured parts from the perspective of engineering application.

The aim of this book is to provide the state of the art of MMM in remanufacturing engineering for students, teachers, researchers and engineers. And the authors sincerely hope that the publication of this book can help readers broaden their knowledge, improve skills and solve problems. It is inevitable to exist errors in this book and criticism and correctness are welcomed. The authors would like to thank postgraduate students Rujun Liu, Jieyan Yao, Shilin Jiang, Cheng Yang, Gang Han, Bin Xiong, Lunwu Zhao, Yuqin Peng, Jie Tang, Jianglong Wang, and Hongmeng Xu for their research on MMM testing in remanufacturing. The authors would also like to acknowledge the financial support by the National Natural Science Foundation of China (50905052, 51675155, 51722502, 52005246 and U20A20295).

Hefei, China   Haihong Huang
December 2020   Zhengchun Qian
                                                                          Zhifeng Liu

Contents

Part I   Introduction to the Metal Magnetic Memory (MMM) Technique

1   Nondestructive Testing for Remanufacturing 3
  1.1   Motivations 3
  1.2   Conventional Nondestructive Testing Techniques 5
  1.3   MMM Technique 5
  1.4   Organization of This Book 7
  References 8

2   Theoretical Foundation of the MMM Technique 9
  2.1   Background 9
  2.2   Microscopic Mechanism 10
  2.3   Macroscopic Theoretical Model 13
      2.3.1   Magnetomechanical Model 13
      2.3.2   Magnetic Charge Model 17
      2.3.3   First Principle Theory 19
  References 23

3   State of the Art of the MMM Technique 25
  3.1   Historical Background 25
  3.2   Theoretical Research 26
  3.3   Experimental Research 27
  3.4   Standard Establishment 30
  3.5   Applications for Remanufacturing 31
  3.6   Problems and Prospects 32
  References 34

Part II   Detection of Damage in Ferromagnetic Remanufacturing Cores by the MMM Technique

4   Stress Induces MMM Signals 39
  4.1   Introduction 39

4.2   Variations in the MMM Signals Induced by Static Stress 40
        4.2.1   Under the Elastic Stage 41
        4.2.2   Under the Plastic Stage 42
        4.2.3   Theoretical Analysis 44
4.3   Variations in the MMM Signals Induced by Cyclic Stress 45
        4.3.1   Under Different Stress Cycle Numbers 46
        4.3.2   Characterization of Fatigue Crack Propagation 49
4.4   Conclusions 52
References 52

5   Frictional Wear Induces MMM Signals 55
    5.1   Introduction 55
    5.2   Reciprocating Sliding Friction Damage 56
            5.2.1   Variations in the Tribology Parameters During
                     Friction 58
            5.2.2   Variations in the Magnetic Memory Signals
                     Parallel to Sliding 60
            5.2.3   Variations in the Magnetic Memory Signals
                     Normal to Sliding 62
            5.2.4   Relationship Between the Tribology
                     Characteristics and Magnetic Signals 65
    5.3   Single Disassembly Friction Damage 66
            5.3.1   Surface Damage and Microstructure Analysis 68
            5.3.2   Variations in the MMM Signals 69
            5.3.3   Damage Evaluation of Disassembly 73
            5.3.4   Verification for Feasibility and Repeatability 76
    5.4   Conclusions 80
    References 81

6   Stress Concentration Impacts on MMM Signals 83
    6.1   Introduction 83
    6.2   Stress Concentration Evaluation Based on the Magnetic
            Dipole Model 84
            6.2.1   Establishment of the Magnetic Dipole Model 84
            6.2.2   Characterization of the Stress Concentration
                     Degree 86
            6.2.3   Contributions of Stress and Discontinuity
                     to MMM Signals 91
    6.3   Stress Concentration Evaluation Based on the Magnetic
            Dual-Dipole Model 95
            6.3.1   Magnetic Scalar Potential 95
            6.3.2   Magnetic Dipole and Its Scalar Potential 97
            6.3.3   Measurement Process and Results 100
            6.3.4   Analysis of the Magnetic Scalar Potential 103
    6.4   Stress Concentration Inversion Method 110

6.4.1   Inversion Model of the Stress Concentration
Based on the Magnetic Source Distribution 110
6.4.2   Inversion of a One-Dimensional Stress
Concentration 112
6.4.3   Inversion of a Two-Dimensional Stress
Concentration 113
6.5   Conclusions 114
References 115

7   Temperature Impacts on MMM Signals 117
7.1   Introduction 117
7.2   Modified J-A Model Based on Thermal and Mechanical
Effects 118
7.2.1   Effect of Static Tensile Stress on the Magnetic Field ... 119
7.2.2   Effect of Temperature on the Magnetic Field 120
7.2.3   Variation in the Magnetic Field Intensity 121
7.3   Measurement of MMM Signals Under Different
Temperatures 122
7.3.1   Material Preparation 122
7.3.2   Testing Method 122
7.4   Variations in MMM Signals with Temperature and Stress 123
7.4.1   Normal Component of the Magnetic Signal 123
7.4.2   Mean Value of the Normal Component
of the Magnetic Signal 125
7.4.3   Variation Mechanism of the Magnetic Signals
Under Different Temperatures 128
7.4.4   Analysis Based on the Proposed Theoretical Model 130
7.5   Conclusions 131
References 132

8   Applied Magnetic Field Strengthens MMM Signals 133
8.1   Introduction 133
8.2   MMM Signal Strengthening Effect Under Fatigue Stress 134
8.2.1   Variations in the MMM Signals with an Applied
Magnetic Field 135
8.2.2   Theoretical Explanation Based on the Magnetic
Dipole Model 137
8.3   MMM Signal Strengthening Effect Under Static Stress 139
8.3.1   Magnetic Signals Excited by the Geomagnetic
Field 140
8.3.2   Magnetic Signals Excited by the Applied
Magnetic Field 142
8.4   Conclusions 146
References 147

**Part III   Evaluation of the Repair Quality of Remanufacturing
                 Samples by the MMM Technique**

9   Characterization of Heat Residual Stress During Repair 151
      9.1   Introduction 151
      9.2   Preparation of Cladding Coating and Measurement
             of MMM Signals 153
             9.2.1   Specimen Preparation 153
             9.2.2   Measurement Method 153
             9.2.3   Data Preprocessing 155
      9.3   Distribution of MMM Signals Near the Heat Affected Zone 156
             9.3.1   Magnetic Signals Parallel to the Cladding Coating 156
             9.3.2   Magnetic Signals Perpendicular to the Cladding
                      Coating 157
             9.3.3   Three-Dimensional Spatial Magnetic Signals 159
             9.3.4   Verification Based on the XRD Method 161
      9.4   Generation Mechanism of MMM Signals in the Heat
             Affected Zone 164
             9.4.1   Microstructure and Phase Transformation 164
             9.4.2   Microhardness Distribution 165
      9.5   Conclusions 166
      References 167

10   Detection of Damage in Remanufactured Coating 169
      10.1   Introduction 169
      10.2   Cladding Coating and Its MMM Measurement 170
      10.3   Result and Discussion 172
              10.3.1   Variations in MMM Signals Under the Fatigue
                       Process 172
              10.3.2   Comparison of the Magnetic Properties
                       from Different Material Layers 174
              10.3.3   Microstructure Analysis 177
      10.4   Conclusions 178
      References 178

11   Detection and Evaluation of Coating Interface Damage 181
      11.1   Introduction 181
      11.2   Theoretical Framework 183
              11.2.1   Fatigue Cohesive Zone Model 183
              11.2.2   Magnetomechanical Model 184
              11.2.3   Numerical Algorithm of the Coupling Model 185
              11.2.4   Calculation of the Magnetic Field Intensity 186
      11.3   Case Analysis for the Theoretical Model 186
              11.3.1   Finite Element Model Setup 186
              11.3.2   Finite Element Simulation Results 188
              11.3.3   Prediction of Interfacial Crack Initiation 190

        11.3.4 Prediction of the Interfacial Crack Propagation
                Behavior 191
    11.4 Experimental Verification 194
        11.4.1 MMM Measurement Method 194
        11.4.2 MMM Signal Analysis 196
        11.4.3 Interfacial Crack Observation 198
    11.5 Conclusions 200
    References 200

Part IV Engineering Applications in Remanufacturing

12 Detection of Damage of the Waste Drive Axle Housing
    and Hydraulic Cylinder 205
    12.1 Introduction 205
    12.2 Application of MMM in the Evaluation of Fatigue Damage
          of the Drive Axle Housing 206
        12.2.1 Relation Between MMM Signals and Fatigue
                Cycles 206
        12.2.2 Relation Between MMM Signals and Deformation
                Degree 209
    12.3 Application of MMM in the Evaluation of Fatigue Damage
          of Retired Hydraulic Cylinders 210
        12.3.1 Threshold Determination Method
                for Remanufacturability Evaluation 210
        12.3.2 Experimental Verification 212
    12.4 Conclusions 215
    References 216

13 Evaluation of the Repair Quality of Remanufactured
    Crankshafts 217
    13.1 Introduction 217
    13.2 Repair Process in Remanufacturing 218
    13.3 Evaluation of the Repair Quality of the Remanufactured
          Coating 219
        13.3.1 Optimization of the Processing Parameters 219
        13.3.2 Effect of the Processing Parameters
                on the Microstructure 221
        13.3.3 Effect of the Processing Parameters
                on the Microhardness 223
        13.3.4 Effect of the Processing Parameters on the Wear
                Resistance 223
    13.4 Repair Quality Evaluation Based on MMM Measurement 226
    13.5 Conclusions 227
    References 228

**14 Development of a High-Precision 3D MMM Signal Testing
   Instrument** 229
  14.1 Introduction 229
  14.2 Framework of the Detection System 230
  14.3 Detailed Processes of Instrument Development 231
     14.3.1 Hardware Design 231
     14.3.2 Software Design 232
  14.4 Calibration of Self-developed Instrument 234
     14.4.1 Static Performance of the Instrument 234
     14.4.2 Ability to React to the Geomagnetic Field 235
  14.5 Testing of the Self-developed Instrument 237
     14.5.1 Testing Method and Process 237
     14.5.2 Display and Analysis of MMM Signals 238
  14.6 Comparison of the MMM Testing Instruments 238
  14.7 Conclusions 240
  References 240

Part I
Introduction to the Metal Magnetic Memory (MMM) Technique

Chapter 1
Nondestructive Testing for Remanufacturing

1.1 Motivations

Global environmental pollution and resource scarcity have become increasingly serious at present. In addition, many pieces of mechanical equipment have been scrapped from various industries, including the automobile, shipping, aerospace and petroleum industries. Taking China as an example, official statistical data from the China Construction Machinery Association in 2014 indicated that product ownership of engineering machinery reached as high as 7 million, 80% of which exceeded warranty. Moreover, the number of registered and scrapped automobiles reached 33.52 million and 8.58 million, respectively, in 2017, as shown in Fig. 1.1 based on the Chinese automobile market investigation report. It can be predicted that the quantity of automobile scrappage will exceed 18 million in the next few years. In addition, there are still 7.28 million engineering machinery, 8 million high-end machine tools, and 1.2 thousand shield tunneling machines in service at present. All of these machinery will be scrapped or upgraded after reaching lifetime limitations, which has great potential for remanufacturing.

Therefore, maximizing the usefulness of waste products has been the key to improving the resource utilization level under this new background. Compared with the products of the traditional manufacturing industry, remanufactured products can reduce costs by 50%, save more than 60% in energy consumption and 70% in materials and reduce emissions by 80%. Taking the first remanufacturing shield tunneling machine in China as an example, the remanufacturing process can save over 20 million yuan, 200 tons of steel, and 260 tons of standard coal while cutting down 700 tons of carbon dioxide.

Consequently, remanufacturing exhibits an affinity to the concepts of sustainable production and sustainable society and has been attracting increasing attention worldwide [1]. For example, the USA has the largest industry scale, exceeding 75 billion dollars [2]. Europe enforces recycling and remanufacturing laws for waste automotive parts, and the European Commission lists remanufacturing as the research plan in "Horizon 2020". Among them, the remanufacturing industry in the UK can create

© Science Press 2021
H. Huang et al., *Metal Magnetic Memory Technique and Its Applications in Remanufacturing*, https://doi.org/10.1007/978-981-16-1590-0_1

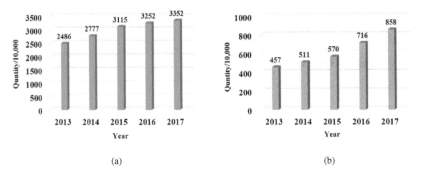

Fig. 1.1 Number variation of automobiles in China: **a** registered automobiles and **b** scrapped automobiles

5 billion pounds of gross national product. The British Lister Petter company can remanufacture 3 thousand disused engines every year. In addition, at least 90% of parts in German automobiles can be reused. In East Asia, Japan exports one-third of its remanufactured products and earns high profits from such practices [3], and China has exerted immense efforts in recent years to promote the sustainable and healthy development of the remanufacturing industry based on the "Made in China 2025" plan.

Remanufacturing, which is an emerging green technology, is defined as a series of processes that allow end-of-life products and parts to be recommercialized as new products with the same quality, functionality and warranty [4]. The main remanufacturing process consists of disassembling, cleaning, inspecting, repairing and reassembling, as shown in Fig. 1.2. Remanufacturing cores disassembled from used products should be inspected before repair. Cores that satisfy remanufacturability requirements will be restored to their original specifications via a series of repair techniques, including laser cladding, plasma spraying, plasma transferred arc welding and electroplating deposition. During the repair process, a cladding coating is prepared by these surface engineering technologies on the damaged surfaces of the cores so that their performance can be improved and the remanufactured parts can remain in service [5]. Therefore, quality control and evaluation of the remanufacturing process is essential because they are directly related to determining whether the reliability of remanufactured parts can meet the standard for new parts. From the perspective of remanufacturing engineering, inspection plays a key role in assessing the damage

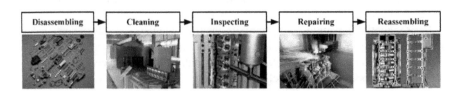

Fig. 1.2 Typical remanufacturing process

degree and repair quality of the cores to ensure that the remanufacturing process proceeds smoothly.

1.2 Conventional Nondestructive Testing Techniques

A nondestructive testing (NDT) technique, which can detect flaws from the interior or surface of a component based on the variations in heat, sound, light, electricity or magnetism attributed to unusual structures and defects, is essential. Compared with traditional destructive testing techniques, NDT has the following advantages: (1) the performance of objects cannot be affected; (2) all the objects can be tested comprehensively rather than sampled in part; and (3) the whole process of manufacturing, service and remanufacturing can be monitored. Therefore, NDT has great importance in petrochemical, aerospace, energy and electricity applications and special equipment.

At present, there are five conventional NDT techniques: penetrant, magnetic particle, radiographic, eddy current and ultrasonic testing. The advantages and disadvantages of these NDT techniques applied to remanufactured products are summarized in Table 1.1. Each NDT technique has its own limitations when it is applied for remanufacturing. For example, penetrant and magnetic particle testing can only find surface defects such as macrocracks and pores instead of inner defects. Radiographic testing is confined to a high cost, a low efficiency and radiation hazards. The eddy current is not applicable for detecting the complicated structure of remanufacturing cores. Ultrasonic testing may not find defects in remanufacturing cladding coatings with columnar crystals and dendrites whose anisotropy affects the propagation of sound waves. Fortunately, metal magnetic memory (MMM) can overcome the weaknesses of the aforementioned conventional NDT techniques when applied to remanufacturing.

1.3 MMM Technique

Ferromagnetic materials are widely used in mechanical structures, particularly in the nuclear, aerospace and military fields, because of their considerable strength, hardness, plasticity and other mechanical properties. During the service process of a mechanical component, ferromagnetic materials exhibit various magnetic properties, such as coercivity, permeability and remanence; they are also sensitive to external applied loads, microstructural features and plastic deformation [6]. The magnetic signals measured from ferromagnetic materials contain a large amount of information about the change in material behaviors. These signals can be used to predict early damage zones, the stress concentration degree and the even residual service life, which determine where and when remanufacturing cores should be repaired.

Table 1.1 Comparison of NDT techniques

Technique	Advantage	Disadvantage	Application
Penetrant	Intuitive results; High sensitivity; Simple process and low cost; Without limitation of material, size and shape	Can only test surface defects; Demand for certain surface quality; Tough environment and heavy pollution	Cannot test internal defects
Magnetic particle	Intuitive results; High sensitivity; Simple process and low cost; Without limitation of size and shape	Can only test surface defects; Demand for certain surface quality; Limitation of orientation; Can only test ferromagnetic materials	Cannot test internal defects
Ultrasonic	High sensitivity; Simple operation and low cost; Wide range of application	Needs couplant; Demand for certain surface quality; Blind area of near-field; Limitation of orientation	Hard to test too thick or too thin objects
Eddy current	Non-contacting; High testing efficiency; Easy automation	Can only test conductive materials; Unsuitable for complicated shapes; Signal can be easily interfered; Can only test surface or near-surface defects	Hard to test thick objects
Radiographic testing	Intuitive results; Wide range of application	High cost; Radiation hazard; Limitation of size	Hard to quantitatively evaluate
Magnetic flux leakage	High testing efficiency; High reliability; Easy automation	Can only test surface defects; Can only test ferromagnetic materials; Unsuitable for complicated shapes	Cannot test internal defects
Metal magnetic memory	High testing efficiency; Evaluate early damage; Without excitation; Without limitation of size and shape	Signal can be easily interfered; Low repeatability and reliability; Hard to quantitatively evaluate	Has potential to test remanufacturing cores and remanufactured products

Most of the remanufacturing cores are made of ferromagnetic materials, as well as cladding coatings composed of iron/cobalt/nickel-based alloys. Accurately determining the damage degree of these cores and/or repair quality is very challenging due to the uncertainty in their service process. Therefore, remanufacturing has its own unique features and specific demands, and it is necessary to find a type of NDT with high sensitivity, easy automation and quantitative characterization [4]. MMM happens to meet these inspection requirements.

As a newly developed NDT technique, MMM is very suitable for online or offline testing due to its simple operation and high efficiency. It can determine the location of stress concentrations, evaluate the early damage degree and predict the residual service life for ferromagnetic materials. Therefore, MMM has become an important technique for evaluating the damage degree of remanufacturing cores and the repair quality of remanufactured coatings. In addition, MMM devices are lightweight and suited for complicated surface detection, which can easily realize flexible detection in remanufacturing. Compared with conventional NDT techniques, MMM has its own unique advantages when applied to remanufacturing.

The fundamental technique of MMM is to use the variation in spontaneous magnetic flux leakage induced by the abnormal internal structure of ferromagnetic materials to determine the defect location and damage degree. Compared with the traditional magnetic flux leakage testing method, MMM can induce spontaneous magnetic signals under the effect of a geomagnetic field without an external excitation device. Therefore, spontaneous magnetic flux leakage signals can effectively evaluate internal defects and early damage. The damaged location can still retain the original spontaneous magnetic flux leakage after the applied load is removed. The type of magnetic signals has the ability to memorize the defect or damage location, and therefore, this NDT technique is also called "metal magnetic memory". Although the intensity of magnetic memory signals is very weak and easily interferes with the environment, they have great potential to be applied to remanufactured products.

1.4 Organization of This Book

This book is divided by four parts. The first part includes Chaps. 1–3, which introduce the fundamentals and the state-of-the-art MMM technique. The second part, including Chaps. 4–8, mainly describes the research of MMM for remanufacturing core damage evaluation. The induction mechanism of magnetic memory signals under various stresses and damage is studied. In addition, the effect of an external magnetic field and temperature environment on magnetic memory signals is also considered. The third part of this book, Chaps. 9–11, refers to MMM for the evaluation of the quality of a remanufacturing repair. Both the performances of the remanufactured coating and its interface are researched in detail. The last part of this book includes Chaps. 12–14. The retired drive axle housing, the damaged hydraulic cylinder, the remanufactured crankshaft, and the high precision three-dimensional

MMM signal testing instruments are introduced to explain the remanufacturing engineering application of MMM.

References

1. M. Matsumoto, S.S. Yang, K. Martinsen et al., Trends and research challenges in remanufacturing. Int. J. Precis. Eng. Manuf.-Green Technol. **3**(1), 129–142 (2016)
2. B.S. Xu, S.Y. Dong, S. Zhu et al., Prospects and development of remanufacturing forming technology. J. Mech. Eng. **48**(15), 96–105 (2012)
3. Y.L. Zhang, H.C. Zhang, J.X. Zhao et al., Review of non-destructive testing for remanufacturing of high-end equipment. J. Mech. Eng. **49**(7), 80–90 (2013)
4. C.M. Lee, W.S. Woo, Y.H. Roh, Remanufacturing: trends and issues. Int. J. Precis. Eng. Manuf.-Green Technol. **4**(1), 113–125 (2017)
5. J.J. Kang, B.S. Xu, H.D. Wang et al., Progress of nondestructive detection technology on remanufacturing coatings. Rare Metal Mater. Eng. **41**(S1), 399–402 (2012)
6. C.C. Li, L.H. Dong, H.D. Wang et al., Current research and development prospects of magnetic non-destructive assessment techniques for fatigue damage. Mater. Rev. **29**(6), 107–113 (2015)

Chapter 2
Theoretical Foundation of the MMM Technique

2.1 Background

The metal magnetic memory (MMM) technique is considered a new nondestructive testing (NDT) method for detecting early damage of ferromagnetic materials or evaluating the repair quality of remanufactured products. However, the research to date on the MMM technique has mainly involved the characteristic extraction of magnetic signals or investigation of their variation trend. This series of studies cannot help us to clarify the generation mechanism of spontaneous magnetic flux leakage and enable the quantitative evaluation of early damage and repair quality based on MMM signals. Therefore, the application of the MMM technique in remanufacturing engineering is seriously restricted.

Only a few researchers have tried to explain how the MMM signal is formed from microscopic magnetic domains. Ren et al. [1] observed variations in the magnetic domains of alloy-20 stainless steel under applied loads. The results indicate that the pattern of magnetic domains in ferromagnetic specimens varies with the stress. When the stress is not concentrated or the stress concentration is small, the magnetic domains in the grains are mainly sheet-like domains, and the domain walls in the same crystal grain are parallel to each other. However, the length and spacing of the domain walls change, and labyrinth domains appear while the stress increases. Furthermore, as the number of labyrinth domains increases, the magnetization at the position of the stress concentration becomes large, and spontaneous magnetic flux leakage forms near the surface.

It is worth noting that some researchers have also been devoted to establishing relevant theoretical models based on physical mechanisms to better describe the variations in MMM signals. Shi et al. [2] considered that the basic principle of the MMM technique is that a ferromagnetic material exhibits a force-magnetic coupling effect in which mechanical energy and magnetic energy are mutually converted. The basic theory of ferromagnetism suggests that the length of ferromagnetic materials placed in an external magnetic field changes due to variations in their magnetization state. That is, ferromagnetic materials have magnetostrictive properties, which is also

© Science Press 2021
H. Huang et al., *Metal Magnetic Memory Technique and Its Applications in Remanufacturing*, https://doi.org/10.1007/978-981-16-1590-0_2

known as the Joule effect. The applied stress changes the orientations of the magnetic domains inside the ferromagnetic material, which alters its magnetic properties. Thus, ferromagnetic materials have an inverse magnetostrictive effect, also known as the Villaiy effect or the magnetomechanical effect, which can help the MMM technique evaluate the residual stress state inside of a ferromagnetic specimen. Wang et al. [3] assumed that a constant distribution of magnetic charge density can be measured near the crack surface, and thus, the MMM signal can also be analyzed using the magnetic dipole method. Leng et al. [4] studied the characteristics of the MMM signal from a V-groove specimen, and the results showed that the MMM signal became nonlinear near the groove, and this nonlinearity increased with the load. In addition, the first principle theory can also be used to illustrate the mechanism for the magnetization of materials from a microscopic perspective. By calculating the relationship among the lattice structure, atomic magnetic moment, and system energy and force, Yang et al. [5] studied how the force influences the magnetic properties of materials and the relationship between the atomic magnetic moment and pressure.

The abovementioned studies related to the mechanism and theory of the MMM technique are both scientifically important and beneficial for engineering applications. Such studies aid the understanding of the distortion phenomena or variation laws of MMM signals in ferromagnetic materials and contribute to the scientific determination of the damage degree or repair quality in remanufacturing. It can be considered that the studies of the microscopic generation mechanism and macroscopic theoretical model for MMM signals in this chapter are the method and application foundation of the MMM technique for Parts 2, 3 and 4 in this book.

2.2 Microscopic Mechanism

In Ref. [6], Qiu et al. assumed that the surface magnetic domain structure of ferromagnetic materials without an external magnetic field in space is shown in Fig. 2.1a. The total areas of the 180° and −180° magnetic domains are equal. When an external magnetic field whose direction is consistent with the 180° magnetic domain is applied to the ferromagnetic materials, the 180° magnetic domains expand while the −180°

Fig. 2.1 The area of the 180° and −180° magnetic domains: **a** without an external magnetic field; **b** with an upward magnetic field; and **c** with a downward magnetic field [6]

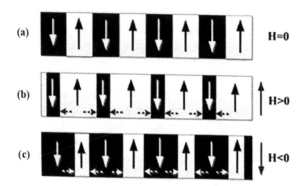

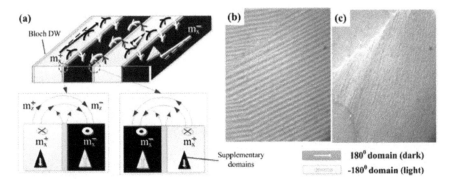

Fig. 2.2 Arrangements of the magnetic domains: **a** schematic diagram; **b** HGO electrical steel in the demagnetized state; and **c** HGO electrical steel at a magnetic field of 880 A m^{-1} [6]

magnetic domains become narrow, as shown in Fig. 2.1b. In contrast, when the direction of the external magnetic field is consistent with the $-180°$ magnetic domain, the areas of the $-180°$ magnetic domains are larger than those of the $180°$ magnetic domains, as shown in Fig. 2.1c.

After the ferromagnetic materials are subjected to stress or an applied magnetic field, there must be a small out-of-plane component of magnetization in the surface materials, which causes magnetic poles and stray fields, as shown in Fig. 2.2a. To study the generation mechanism of MMM signals, the dynamic behaviors of magnetic domains in high permeability grain oriented (HGO) electrical steel are observed in situ by using the magneto-optical Kerr effect [6]. The magnetic domain patterns of HGO electrical steel in demagnetized and magnetized states are shown in Fig. 2.2b and c, respectively. An obvious grain boundary distribution as well as magnetic domain texture can be simultaneously observed. The grain boundary separates two adjoining grains with different crystal orientations. Both the width and orientation of the magnetic domains vary with each grain. The magnetic field leakage around the grain boundary is substantially higher than that in the grain interior when HGO electrical steel is magnetized. The larger stray field around the grain boundary induces a larger MMM signal, which can be reflected by the brighter or darker intensity in the magnetic domain pattern.

The surface magnetic domain patterns under different magnetic field excitations and stress effects are shown in Figs. 2.3, 2.4 and 2.5. Under the effect of a magnetic field of 177 A m^{-1}, all the surface magnetic domains of HGO electrical steel orientate in the $180°$ direction when the applied stress is 0 MPa or 30.9 MPa, as shown in Figs. 2.3a and 2.4a. Only a few $-180°$ domains appear when the stress increases to 61.9 MPa as shown in Fig. 2.5a. As the excitation magnetic field strength decreases to 25 A m^{-1}, all the magnetic domains are still retained in the $180°$ direction when the applied stress is 0 MPa, as shown in Fig. 2.3b. However, many $-180°$ domains are observed under a stress of 30.9 MPa, and the areas of the $-180°$ domains expand when the stress increases to 61.9 MPa as shown in Fig. 2.4b and Fig. 2.5b, respectively. When the magnetic field strength decreases to 0 A m^{-1}, $-180°$ magnetic domains

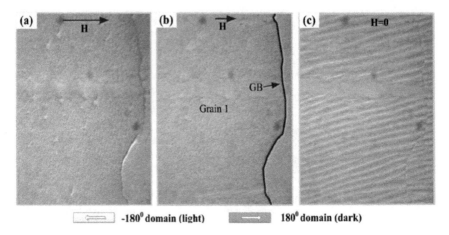

-180⁰ domain (light)   180⁰ domain (dark)

Fig. 2.3 Patterns of the magnetic domains under a stress of 0 MPa: **a** at a magnetic field of 177 A m^{-1}; **b** at a magnetic field of 25 A m^{-1}; and **c** at a magnetic field of 0 A m^{-1} [6]

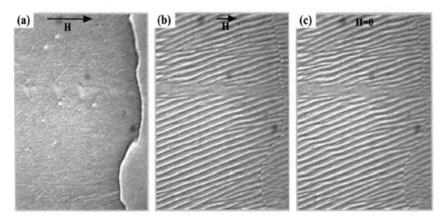

Fig. 2.4 Patterns of the magnetic domains under a stress of 30.9 MPa: **a** at a magnetic field of 177 A m^{-1}; **b** at a magnetic field of 25 A m^{-1}; and **c** at a magnetic field of 0 A m^{-1} [6]

also appear at a stress of 0 MPa, as shown in Fig. 2.3c. In addition, the $-180°$ magnetic domains are expanded, and the $180°$ magnetic domains are narrowed continuously at a stress of 30.9 MPa and 61.9 MPa, respectively, which can be seen in Fig. 2.4c and Fig. 2.5c. Under the effect of applied stress, the demagnetization effect hinders $180°$ grain boundary movement. Therefore, the higher excitation magnetic field strength can lead to the completion of all $180°$ grain boundary movement. The results indicate to some extent that the microscopic generation mechanism of MMM signals is the change in the magnetic domains inside the material caused by the magnetomechanical effect [6].

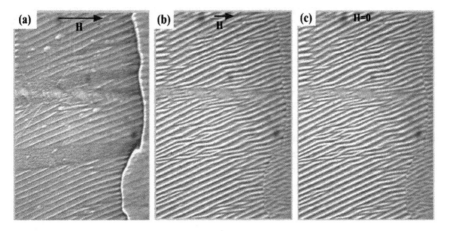

Fig. 2.5 Patterns of the magnetic domains under a stress of 61.9 MPa: **a** at a magnetic field of 177 A m^{-1}; **b** at a magnetic field of 25 A m^{-1}; and **c** at a magnetic field of 0 A m^{-1} [6]

2.3 Macroscopic Theoretical Model

2.3.1 *Magnetomechanical Model*

Based on observations of microscopic magnetic domains, it can be found that when ferromagnetic materials in an external magnetic environment are subjected to stress, their magnetic properties change due to the magnetomechanical effect. In Ref. [7], Jiles et al. established a classical magnetomechanical model according to the thermo-dynamic relations and the approach law of irreversible magnetization. Considering an ideal condition of an anhysteretic relation between stress and magnetization first, an applied uniaxial stress σ acts in some respects like an applied magnetic field operating through the magnetostriction λ. The additional field H can be described by considering the energy A of the system along the reversible anhysteretic magnetization curve:

$$A = \mu_0 H M + \frac{\mu_0}{2}\alpha M^2 + \frac{3}{2}\sigma\lambda + TS \qquad (2.1)$$

where T is the temperature, S is the entropy and $\mu_0 \alpha M^2/2$ is the self-coupling energy. The dimensionless term α represents the coupling strength of the individual magnetic moments to the magnetization M. The effective magnetic field causes a change in magnetization and therefore is determined by the derivative of this energy with respect to magnetization M. The derivative of entropy with respect to bulk magnetization M in a ferromagnet will be negligible in the cases under consideration because the fields applied here do not increase the ordering within the domain. Therefore, the effective field is given by:

$$H_e = \frac{1}{\mu_0} \frac{dA}{dM} = H + \alpha M + \frac{3\sigma}{2\mu_0} \frac{d\lambda}{dM} \tag{2.2}$$

where the effects of stress have been incorporated into the equivalent effective field. It is therefore implicit in this description of the theory that the anhysteretic magnetization under field H and stress σ is identical to the anhysteretic magnetization under an equivalent effective magnetic field. This relationship can be determined if the magnetostriction coefficients λ_{100} and λ_{111} of a certain domain configuration are known. However, in practice, this domain configuration in a material cannot be known in advance. Therefore, it is necessary to develop an empirical model to describe the relation between bulk magnetostriction and bulk magnetization. Since the magnetization must be symmetric about $M = 0$, a simple series expansion gives:

$$\lambda = \sum_{i=0}^{\infty} \gamma_i(\sigma) M^{2i} \tag{2.3}$$

where γ_i is the coefficient related to the material. A reasonable first approximation of the magnetostriction of ferromagnetic materials can be obtained by including the terms up to $i = 2$:

$$\lambda = \gamma_1(\sigma) M^2 + \gamma_2(\sigma) M^4 \tag{2.4}$$

$$\gamma_i(\sigma) = \gamma_i(0) + \sum_{n=1}^{\infty} \frac{\sigma^n}{n!} \gamma_i^n(0) \tag{2.5}$$

where $_i^n(0)$ is the nth derivative of γ_i with respect to the stress at $\sigma = 0$. The approximative magnetostriction can be accepted when $n = 1$:

$$\lambda = \left(\gamma_1(0) + \gamma_1'(0)\sigma \right) M^2 + \left(\gamma_2(0) + \gamma_2'(0)\sigma \right) M^4 \tag{2.6}$$

For isotropic materials, the stress dependence of the anhysteretic magnetization curve can be determined from the equation:

$$M_{an}(H, \sigma) = M_s \left[\coth\left(\frac{H_e}{a} \right) - \frac{a}{H_e} \right]$$
$$= M_s \left[\coth\left(\frac{H + \alpha M + H_\sigma}{a} \right) - \frac{a}{H + \alpha M + H_\sigma} \right] \tag{2.7}$$

where $a = k_B T / \mu_0 M$ determines the shape of the anhysteretic magnetization curve, k_B is the Boltzmann constant, μ_0 is the vacuum permeability and M_s is the saturation magnetization. Selecting the relevant parameter values in Ref. [7], the anhysteretic magnetization M_{an} varying with stress σ is plotted in Fig. 2.6 based on Eq. (2.7). It can be seen that the anhysteretic magnetization increases first and then decreases

Fig. 2.6 Anhysteretic
stress-magnetization curve

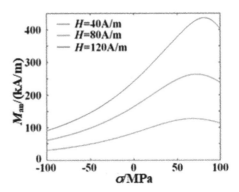

with increasing stress. Under the same stress level, the anhysteretic magnetization increases with the increase in the magnetic field of the external environment.

The ideal phenomenon of anhysteretic magnetization cannot appear; thus, the energy loss caused by the domain pinning effect should be considered. The total magnetization M can be considered to be composed of the reversible magnetization M_{rev} and irreversible magnetization M_{irr}:

$$M = M_{rev} + M_{irr} \tag{2.8}$$

where the reversible magnetization component is contributed by domain wall bending, while the irreversible magnetization component is induced by the irreversible movement of the domain wall. Thus, we can obtain:

$$M_{rev} = c(M_{an} - M_{irr}) \tag{2.9}$$

where c describes the flexibility of the magnetic domain walls. This equation can then be differentiated with respect to the elastic energy W supplied to the material as a result of the applied stress:

$$\frac{dM_{rev}}{dW} = c\left(\frac{dM_{an}}{dW} - \frac{dM_{irr}}{dW}\right) \tag{2.10}$$

To describe the hysteretic behavior of the magnetization curve, it is assumed that the derivative of the irreversible magnetization with respect to the elastic energy follows the approach law:

$$\frac{dM_{irr}}{dW} = \frac{1}{\xi}(M_{an} - M_{irr}) \tag{2.11}$$

where ξ is a coefficient with dimensions of energy per unit volume. Then, the derivative of the total magnetization with respect to the elastic energy is obtained:

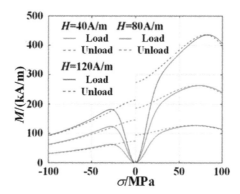

Fig. 2.7 Stress-magnetization curve

$$\frac{dM}{dW} = \frac{1-c}{\xi}(M_{an} - M_{irr}) + c\frac{dM_{an}}{dW} \tag{2.12}$$

Because the differential of the elastic energy is $dW = (\sigma/E)d\sigma$, the final stress-magnetization constitutive relationship can be obtained:

$$\frac{dM}{d\sigma} = \frac{\sigma}{E_a\xi}(M_{an} - M) + c\frac{dM_{an}}{d\sigma} \tag{2.13}$$

In the same way, selecting the relevant parameter values in Ref. [7], the magnetization M varies with stress σ during the loading and unloading process and is plotted in Fig. 2.7 based on Eq. (2.13). The overall magnetization also increases with the increase in the external environment magnetic field. However, the magnetization path during the unloading process cannot vary along the original anhysteretic magnetization due to the effect of domain pinning and irreversible magnetization. In addition, the magnetization curve during the loading process is also different from that of the unloading process.

Although the classical Jiles magnetomechanical model can solve the relationship between magnetization and stress to some extent, some researchers have also proposed modified models to increase its accuracy and applicability. Shi et al. [8] established a nonlinear coupled model to improve the quantitative evaluation of the magnetomechanical effect. Excellent agreement has been achieved between the predictions from the present model and previous experimental results. In comparison with Jiles' model, the prediction accuracy is improved greatly by the present model, particularly for the compression case. Xu et al. [9] constructed a modified J-A model to describe the magnetomechanical effect in the elastic stress stage, which can simulate the variation in MMM signals to become steadier. These theoretical calculation results indicate that under a weak magnetic field environment such as the geomagnetic field, the change in stress causes a change in the magnetization of the material. The magnetomechanical coupling effect is considered to be the basic principle of MMM signal formation.

2.3.2 Magnetic Charge Model

For the detection of defects in ferromagnetic materials, the MMM technique has something in common with the magnetic flux leakage technique. Based on Maxwell's magnetostatic equations and the boundary conditions, the strength of the magnetic field $\mathbf{H}_m$ at a space point caused by the magnetization $\mathbf{M}$ of the ferromagnetic material can be described by:

$$\mathbf{H}_m(\mathbf{r}) = \frac{1}{4\pi}\int_{\forall} \frac{-\nabla \cdot \mathbf{M}(\mathbf{s})}{|\mathbf{r} - \mathbf{s}|^3}(\mathbf{r} - \mathbf{s})d\forall + \frac{1}{4\pi}\int_{S} \frac{\mathbf{n} \cdot \mathbf{M}(\mathbf{s})}{|\mathbf{r} - \mathbf{s}|^3}(\mathbf{r} - \mathbf{s})dS \qquad (2.14)$$

where $\forall$ and S are the volume and surface area of the ferromagnetic material, respectively, $\mathbf{s}$ is the location vector of any point inside the ferromagnetic material or on its surface, $\mathbf{r}$ is the distance between the charge element and space point, and $\mathbf{n}$ is the outside unit vector normal to the surface of the ferromagnetic material.

Based on different situations and requirements, in Ref. [10], Shi et al. considered that the stress concentration zone can be simplified to models of 1D lines, 2D rectangles and 3D bodies, as shown in Fig. 2.8. The sizes of the length, width or depth of the defects are also marked in the figure. Due to the local stress concentration, the

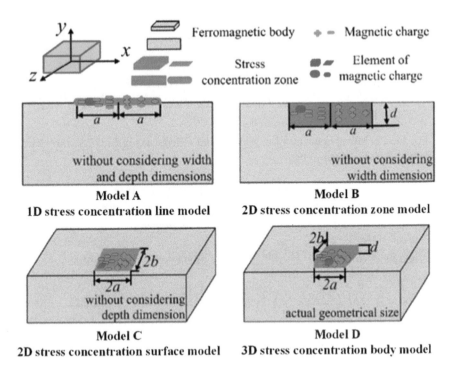

Fig. 2.8 Different magnetic charge models for the local stress concentration [10]

plastic deformation in the defect zone reaches a maximum at the center position and linearly decreases to zero at the edge. Generally, the effect of plastic deformation on magnetic properties is attributed to the generation of dislocation density. According to the theory of magnetic charge, dislocations will lead to the accumulation of magnetic charge in the defect zone. For simplicity, a linear functional relationship is assumed for the plastic strain and magnetic charge density.

First, it is assumed that the charge density for the 1D stress concentration line model (Model A) in Fig. 2.8 is described by:

$$\rho(m, n) = \begin{cases} (m + a)\rho_1/a + \rho_0 & m \in [-a, 0) \\ (m - a)\rho_1/a - \rho_0 & m \in (0, a] \\ 0 & \text{elsewhere} \end{cases} \quad (2.15)$$

where $\rho_1 + \rho_0$ is the maximum magnetic charge density and ρ_0 is the minimum magnetic charge density in the stress concentration line model. m denotes the value for the x-axis in the inner zone of the local stress concentration line model. In addition, the strength of the magnetic field $\mathbf{H}_m$ at a space point caused by the charge elements can be simplified as:

$$\mathbf{H}_m(\mathbf{r}) = \frac{1}{2\pi} \int_S \frac{\rho(\mathbf{s})}{|\mathbf{r} - \mathbf{s}|^2}(\mathbf{r} - \mathbf{s})dS \quad (2.16)$$

Hence, the magnetic field $\mathbf{H}_m$ can be rewritten as:

$$\mathbf{H}_m(\mathbf{r}) = \frac{\rho_1}{2\pi} \int_{-a}^{0} \left(\frac{m + a}{a} + \frac{\rho_0}{\rho_1} \right) \frac{\mathbf{r}}{|\mathbf{r}|^2} dm + \frac{\rho_1}{2\pi} \int_{0}^{a} \left(\frac{m - a}{a} - \frac{\rho_0}{\rho_1} \right) \frac{\mathbf{r}}{|\mathbf{r}|^2} dm \quad (2.17)$$

For the 2D stress concentration zone model (Model B) in Fig. 2.8, the charge density can be assumed to be:

$$\rho(m, n) = \begin{cases} (m + a)\rho_1/a + \rho_0 & m \in [-a, 0), n \in [-d, 0] \\ (m - a)\rho_1/a - \rho_0 & m \in (0, a], n \in [-d, 0] \\ 0 & \text{elsewhere} \end{cases} \quad (2.18)$$

Then, the magnetic field $\mathbf{H}_m$ can also be simplified and rewritten as:

$$\mathbf{H}_m(\mathbf{r}) = \frac{\rho_1}{2\pi} \int_{-a}^{0} \int_{-d}^{0} \left(\frac{m + a}{a} + \frac{\rho_0}{\rho_1} \right) \frac{\mathbf{r}}{|\mathbf{r}|^2} dm dn + \frac{\rho_1}{2\pi} \int_{0}^{a} \int_{-d}^{0} \left(\frac{m - a}{a} - \frac{\rho_0}{\rho_1} \right) \frac{\mathbf{r}}{|\mathbf{r}|^2} dm dn \quad (2.19)$$

Similarly, for the 2D stress concentration surface model (Model C) in Fig. 2.8, the charge density can be expressed as:

$$\rho(m, t) = \begin{cases} (m+a)\rho_1/a + \rho_0 & m \in [-a, 0), t \in [-b, b] \\ (m-a)\rho_1/a - \rho_0 & m \in (0, a], t \in [-b, b] \\ 0 & \text{elsewhere} \end{cases} \quad (2.20)$$

The magnetic field $\mathbf{H}_m$ can be written as:

$$\mathbf{H}_m(\mathbf{r}) = \frac{\rho_1}{4\pi} \int\limits_{-a}^{0} \int\limits_{-b}^{b} \left(\frac{m+a}{a} + \frac{\rho_0}{\rho_1}\right) \frac{\mathbf{r}}{|\mathbf{r}|^3} dmdt + \frac{\rho_1}{4\pi} \int\limits_{0}^{a} \int\limits_{-b}^{b} \left(\frac{m-a}{a} - \frac{\rho_0}{\rho_1}\right) \frac{\mathbf{r}}{|\mathbf{r}|^3} dmdt \quad (2.21)$$

Finally, the charge density of the 3D stress concentration body model (Model D) in Fig. 2.8 can be described as:

$$\rho(m, n, t) = \begin{cases} (m+a)\rho_1/a + \rho_0 & m \in [-a, 0), n \in [-d, 0], t \in [-b, b] \\ (m-a)\rho_1/a - \rho_0 & m \in (0, a], n \in [-d, 0], t \in [-b, b] \\ 0 & \text{elsewhere} \end{cases} \quad (2.22)$$

And the magnetic field $\mathbf{H}_m$ can be obtained as:

$$\mathbf{H}_m(\mathbf{r}) = \frac{\rho_1}{4\pi} \int\limits_{-a}^{0} \int\limits_{-d}^{0} \int\limits_{-b}^{b} \left(\frac{m+a}{a} + \frac{\rho_0}{\rho_1}\right) \frac{\mathbf{r}}{|\mathbf{r}|^3} dmdndt$$

$$+ \frac{\rho_1}{4\pi} \int\limits_{0}^{a} \int\limits_{-d}^{0} \int\limits_{-b}^{b} \left(\frac{m-a}{a} - \frac{\rho_0}{\rho_1}\right) \frac{\mathbf{r}}{|\mathbf{r}|^3} dmdndt \quad (2.23)$$

Based on these magnetic charge models, a comparison between the theoretical calculation results and experimental data is presented. With the simulation parameters in Ref. [10], the axial component of the MMM signal is shown for all four theoretical models in Fig. 2.9. The simulation results obtained using these magnetic charge models are all very consistent with the experimental results, thereby confirming the effectiveness of these magnetic charge models for simulating stress concentrations, defects and other damage.

2.3.3 First Principle Theory

In Ref. [11–12], Liu et al. considered that the generation mechanism of MMM signals can also be explained by first principle calculations based on density function theory with pseudopotential and plane-wave methods. A one-to-one correspondence relationship exists between the stress potential, $v\left(\vec{r}\right)$, and the electronic density function, $\rho\left(\vec{r}\right)$, and the electronic kinetic energy T and potential energy U inside the system

Fig. 2.9 Comparison of the axial component of the MMM signals between the theoretical models and experimental data [10]

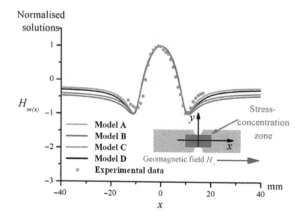

are the only functions of $\rho\left(\vec{r}\right)$. Then, the energy functional of the whole system in the ground state can be expressed as:

$$E\left[\rho\left(\vec{r}\right)\right] = \int v\left(\vec{r}\right)\rho\left(\vec{r}\right)d\vec{r} + F\left[\rho\left(\vec{r}\right)\right] = \int v\left(\vec{r}\right)\rho\left(\vec{r}\right)d\vec{r} + \int \psi^*(T+U)\psi dr \tag{2.24}$$

where $\vec{r}$ is the electron coordinates, $F\left[\rho\left(\vec{r}\right)\right]$ is a universal function that is applicable to any particle system and any external potential, and ψ is the electron wave function of the multiparticle system. In the multiparticle system, the specific form of $F\left[\rho\left(\vec{r}\right)\right]$ is difficult to obtain, which includes both classical Coulomb energy and the exchange interaction energy among the electronic systems. Therefore, we consider that $F\left[\rho\left(\vec{r}\right)\right]$ is the sum of the kinetic energy part $G\left[\rho\left(\vec{r}\right)\right]$ and the classic potential energy part $U\left[\rho\left(\vec{r}\right)\right]$. A kinetic energy functional $T\left[\rho\left(\vec{r}\right)\right]$ with interacting particles is replaced by a known kinetic energy functional $T_s\left[\rho\left(\vec{r}\right)\right]$ without interacting particles. Both T and T_s have the same density function, and the parts that cannot be converted are contained in the exchange correlation potential $E_{xc}\left[\rho\left(\vec{r}\right)\right]$. Then, the universal function, kinetic energy part, and classic potential energy part can be expressed as:

$$F\left[\rho\left(\vec{r}\right)\right] = G\left[\rho\left(\vec{r}\right)\right] + \frac{1}{2}\iint \frac{\rho\left(\vec{r}\right)\rho\left(\vec{r}'\right)}{\left|\vec{r}-\vec{r}'\right|}d\vec{r}d\vec{r}' \tag{2.25}$$

$$G\left[\rho\left(\vec{r}\right)\right] = T_s\left[\rho\left(\vec{r}\right)\right] + E_{xc}\left[\rho\left(\vec{r}\right)\right] \tag{2.26}$$

$$U\left[\rho\left(\vec{r}\right)\right] = \frac{1}{2}\iint \frac{\rho\left(\vec{r}\right)\rho\left(\vec{r'}\right)}{\left|\vec{r}-\vec{r'}\right|} d\vec{r}d\vec{r'} \tag{2.27}$$

To structure $T_s\left[\rho\left(\vec{r}\right)\right]$, N single-particle wave functions $\varphi_i\left(\vec{r}\right)$ ($i = 1, 2, ...$) are introduced, and the electronic density function and the noninteracting kinetic energy part functional can be shown as:

$$\rho\left(\vec{r}\right) = \sum_{i=1}^{N}\left|\varphi_i\left(\vec{r}\right)\right|^2 \tag{2.28}$$

$$T_s\left[\rho\left(\vec{r}\right)\right] = \sum_{i=1}^{N}\int d\vec{r}\varphi_i^*\left(\vec{r}\right)(-\nabla^2)\varphi_i\left(\vec{r}\right) \tag{2.29}$$

Therefore, the ground state energy functional of the system is:

$$E\left[\rho\left(\vec{r}\right)\right] = \int v\left(\vec{r}\right)\rho\left(\vec{r}\right)d\vec{r} + \frac{1}{2}\iint \frac{\rho\left(\vec{r}\right)\rho\left(\vec{r'}\right)}{\left|\vec{r}-\vec{r'}\right|} d\vec{r}d\vec{r'}$$

$$+ \sum_{i=1}^{N}\int d\vec{r}\varphi_i^*\left(\vec{r}\right)(-\nabla^2)\varphi_i\left(\vec{r}\right) + E_{xc}\left[\rho\left(\vec{r}\right)\right] \tag{2.30}$$

The best form of the energy functional under the single-electron states is obtained through the variation of the system ground state energy $E\left[\rho\left(\vec{r}\right)\right]$ with respect to the density function $\rho\left(\vec{r}\right)$:

$$\frac{\delta\left\{E\left[\rho\left(\vec{r}\right)\right]\right\}}{\delta\rho\left(\vec{r}\right)} = 0 \tag{2.31}$$

Therefore, the single electron equation, which is used to solve the multiparticle system, can then be expressed as:

$$\left[-\frac{1}{2}\nabla^2 + V_{eff}\left[\rho\left(\vec{r}\right)\right]\right]\varphi_i\left(\vec{r}\right) = \varepsilon_i\varphi_i\left(\vec{r}\right) \tag{2.32}$$

where ε_i is the Kohn–Sham eigenvalue and $V_{eff}\left(\vec{r}\right)$ denotes the effective potential. Then, the energy of particles in an effective potential field can be expressed as:

$$V_{eff}\left(\vec{r}\right) = \upsilon\left(\vec{r}\right) + \frac{\delta E_{xc}\left[\rho\left(\vec{r}\right)\right]}{\delta\rho\left(\vec{r}\right)} + \int d\vec{r}' \frac{\rho\left(\vec{r}\right)}{\left|\vec{r} - \vec{r}'\right|} \tag{2.33}$$

In ferromagnetic materials, the atomic magnetic moment is derived from the electron spin under the shell, and electron orbital motion in the ground state has no contribution to the magnetic field. The surplus magnetic moment of electrons in the complete shell of the atom is zero and makes no contribution to the magnetic characteristics of the material. Only electrons in the incomplete shell can produce an atomic magnetic moment. According to the Stoner criterion, the magnetization of the solid is defined as

$$M = N\mu_B\left[\rho(E)_\uparrow - \rho(E)_\downarrow\right] \tag{2.34}$$

where N is the number of electrons and μ_B denotes the atomic magnetic moments. Finally, the MMM signals can be described by:

$$B = \mu_0(H + M) = \mu_0\left\{H + N\mu_B\left[\rho(E)_\uparrow - \rho(E)_\downarrow\right]\right\} \tag{2.35}$$

where μ_0 denotes the magnetic conductivity in a vacuum, and H expresses the magnetic field strength. Therefore, the magnetomechanical effect of the MMM technique can be researched by calculating the system energy and electron density of the state distribution based on density function theory.

Based on the first principle theory, Yang et al. [5] explored the mechanism for the magnetization of materials from a microscopic perspective. By calculating the relationship among the lattice structure, atomic magnetic moment, and system energy and force, they studied how the force influences the magnetic properties of materials and the relationship between the atomic magnetic moment and pressure. Liu et al.

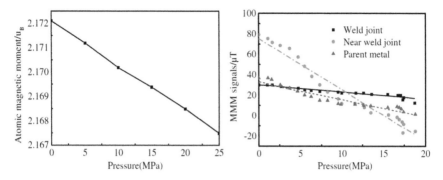

Fig. 2.10 Phenomenon of MMM signals decreasing with stress: **a** variation of the atomic magnetic moment with pressure; and **b** MMM signal distribution [11, 12]

[11, 12] also studied the quantitative relationship between the stress concentration and MMM signals by the first principle theory and analyzed how material doping influences MMM signals. Figure 2.10 shows that both the theoretical simulation results and the experimental data reflect the linearly decreasing trend of MMM signals with increasing stress.

References

1. J.L. Ren, C. Chen, C.K. Liu et al., Experimental research on microcosmic mechanism of stress-magnetic effect for magnetic memory testing, in *17th World Conference on Nondestructive Testing*, Shanghai, vol. 28, no. 5 (2008), pp. 41–44
2. P.P. Shi, S.Q. Su, Z.M. Chen, Overview of researches on the nondestructive testing method of metal magnetic memory: status and challenges. J. Nondestruct. Eval. **39**, 43 (2020)
3. Z.D. Wang, K. Yao, B. Deng et al., Theoretical studies of metal magnetic memory technique on magnetic flux leakage signals. NDT&E Int. **43**(4), 354–359 (2010)
4. J.C. Leng, M.Q. Xu, J.W. Li et al., Characterization of the elastic-plastic region based on magnetic memory effect. Chin. J. Mech. Eng. **23**(4), 532–536 (2010)
5. L.J. Yang, B. Liu, L.J. Chen et al., The quantitative interpretation by measurement using the magnetic memory method (MMM)-based on density functional theory. NDT&E Int. **55**, 15–20 (2013)
6. F.S. Qiu, W. Ren, G.Y. Tian et al., Characterization of applied tensile stress using domain wall dynamic behavior of grain-oriented electrical steel. J. Magn. Magn. Mater. **432**, 250–259 (2017)
7. D.C. Jiles, Theory of the magnetomechanical effect. J. Phys. D Appl. Phys. **28**(8), 1537–1546 (1995)
8. P.P. Shi, K. Jin, X.J. Zheng, A general nonlinear magnetomechanical model for ferromagnetic materials under a constant weak magnetic field. J. Appl. Phys. **119**, (2016)
9. M.X. Xu, M.Q. Xu, J.W. Li et al., Using modified J-A model in MMM detection at elastic stress stage. Nondestruct. Testi. Eval. **27**(2), 121–138 (2012)
10. P.P. Shi, X.J. Zheng, Magnetic charge model for 3D MMM signals. Nondestruct. Test. Eval. **31**(1), 45–60 (2016)

11. B. Liu, Y. Fu, R. Jian, Modelling and analysis of magnetic memory testing method based on the density functional theory. Nondestruct. Test. Eval. **30**(1), 13–25 (2015)
12. B. Liu, Y. Fu, B. Xu, Study on metal magnetic memory testing mechanism. Res. Nondestruct. Eval. **26**, 1–12 (2015)

Chapter 3
State of the Art of the MMM Technique

3.1 Historical Background

In 1975, Misra found a transient magnetic field during crack propagation under low carbon steel tensile experiments [1]. This result showed that spontaneous magnetic flux leakage was formed when the ferromagnetic materials were subjected to stress. Based on this, the concept of metal magnetic memory (MMM) was first proposed by Russian scholar Doubov in 1994, and MMM was applied in the nondestructive testing (NDT) field [2]. In 1995, Jiles from Iowa State University introduced the approach law with the anhysteretic magnetization process and discussed the relationship among the stress, magnetization and magnetostriction. Then, the Jiles magnetomechanical model was established, which promoted the development of MMM [3]. At the 50th International Welding Conference in 1998, Doubov proposed the theory of the "metal stress concentration zone-metal microstructure change-magnetic memory effect", which received great attention worldwide [4]. In 2000, Ren et al. [5] published a book about MMM and pointed out that the physical mechanism of MMM was based on the magnetomechanical effect of force/magnetism coupling and the magnetoelastic effect. The micromechanism of MMM has been recognized more clearly since entering the 21st century. At the initial stage, the magnetic domains distribute randomly, and no magnetization appears on the macroscale. When the ferromagnetic materials are affected by the geomagnetic field, the magnetic domains rotate and move partially, and the overall zone displays weak magnetization. When ferromagnetic materials are subjected to the combined actions of the geomagnetic field and the load, tension tends to orient the domains in the direction of the applied load for a positive magnetostrictive material and compression orients the domains perpendicular to the loading direction due to the piezomagnetic effect [6], as shown in Fig. 3.1a. The magnetic domain movements induce variations in the spontaneous magnetization and permeability. Consequently, a leakage magnetic field is generated in the stress concentration zones [7], where tangential components reach the maximum value and normal components change polarity and have a value of zero [6], as shown in Fig. 3.1b. Magnetic signals are retained even if the load is removed

© Science Press 2021
H. Huang et al., *Metal Magnetic Memory Technique and Its Applications in Remanufacturing*, https://doi.org/10.1007/978-981-16-1590-0_3

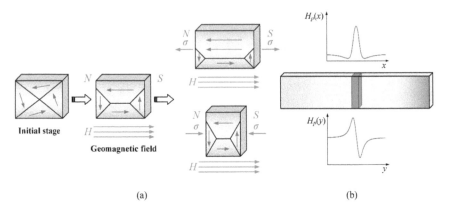

Fig. 3.1 Schematic diagram of MMM: **a** movement of the magnetic domains and **b** distribution of magnetic signals in the stress concentration zone

because of the irreversible rotation of magnetic domains caused by stress concentrations; thus, the signals can apparently memorize the locations of flaws and stress concentrations in ferromagnetic materials [8].

3.2   Theoretical Research

Four theoretical models are studied to explain the generation mechanism of magnetic memory signals in the aforementioned fundamental experiments, as shown in Table 3.1 and Fig. 3.2. First, the distribution of magnetic memory signals induced by stress concentrations was achieved via finite element simulation of force/magnetism coupling [9, 10]. Singh et al. [10] found that geometry and plastic deformation substantially influenced the magnetic signals, as shown in Fig. 3.2a. The Jiles magnetomechanical model based on magnetic domain wall motion theory was studied and modified to further explain the magnetomechanical effect [11, 12]. The hysteresis loop was calculated through hysteresis loss and the energy conservation law. Excellent agreement was achieved between the predictions from the modified magnetomechanical model and the experimental results, as shown in Fig. 3.2b. In Fig. 3.2c, magnetic charge planes with the same magnetic charge density and opposite magnetic poles accumulated on either side of the defect zone when dislocation occurred due to stress concentrations. Wang et al. [13, 14] researched the magnetic dipole model in detail and quantified the relationship between magnetic memory signals and the depth, width and location of defects. This model was successfully applied to the monitoring of fatigue crack propagation [15, 16]. Shi et al. [17] modified the magnetic dipole model and derived the magnetic charge model from 2D to 3D stress concentration zones to extend the application range. From the perspective of the microstructure, lattice distortion can also lead to spontaneous magnetic flux leakage signals. Liu et al. [18, 19] calculated the atomic magnetic moment in the spin polarization system based

Table 3.1 Theoretical model of MMM

Model	Step	Mechanism
Force/magnetism coupling	(1) Modelling and meshing; (2) Adding constraints and loads in statics; (3) Stress is a function of permeability; (4) Calculating spontaneous magnetic flux leakage in magnetostatics	Stress induces permeability, which affects the magnetic flux leakage distribution
Magnetomechanical model	(1) Langevin function describes anhysteretic magnetization curves; (2) Calculating stress magnetization curves based on the domain wall pinning effect and approach law	Stress magnetization constitutive relation based on magnetic domain movement theory
Magnetic dipole model	(1) Magnetic charge planes with the same magnetic charge density and opposite magnetic poles accumulate on either side of the defect zone; (2) Surface magnetic flux leakage is calculated by integrating any point in space	Magnetic charges accumulate at the defect zone under the effect of dislocation induced by plastic deformation
Quantum theory	(1) Atomic magnetic moments are arranged in parallel directions, which shows ferromagnetism; (2) Solving Schrodinger equation and calculating system energy	Stress induces lattice distortion, which leads to the spontaneous magnetic flux leakage signals

on quantum theory. The result is shown in Fig. 3.2d, which indicates that the stress and magnetic signals exhibit a good linear relationship.

3.3 Experimental Research

Due to the aforementioned advantages, MMM can play an important role in remanufacturing core damage assessment and repair quality evaluation. Experts and scholars have conducted considerable research in this area. In the fundamental experiments, five factors that can affect the variation in magnetic memory signals are discussed as follows:

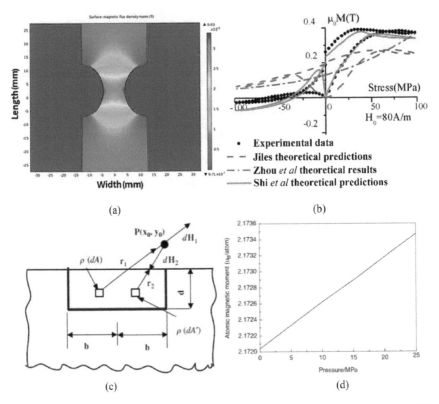

(a)   (b)

(c)   (d)

Fig. 3.2 Theoretical models and calculation results for MMM: **a** finite element simulation [10]; **b** modified magnetomechanical model [11]; **c** magnetic dipole model [12]; and **d** quantum theory [19]

- **Fatigue cycles**: During the service process, cyclic loading directly affects the residual life of ferromagnetic materials. Thus, specimens of 18CrNi4A steel [20], 45CrNiMoVA steel [21] and 45 steel [22], as common ferromagnetic materials used in the remanufacturing industry, were tested under fatigue tensile loads by a team from the Academy of Armored Forces Engineering in China. The quantitative relationship between the magnetic signal characteristic values and fatigue cycle number was presented, as shown in Fig. 3.3.
- **Crack length**: Crack initiation will occur after certain fatigue cycles, and crack propagation will also influence magnetic memory signals. A research group from Hefei University of Technology in China investigated the magnetic signals induced by fatigue bending loads. The results showed that the maximum gradient of the normal component of the magnetic signal K_{max} increased linearly with an increase in the crack length in Q345 structural steel [23], whereas it increased exponentially in the welded joint of Q235 steel [24], as shown in Fig. 3.4.
- **Plastic deformation**: The properties of ferromagnetic materials will change considerably when they enter the plastic stage. The effect of plastic deformation on

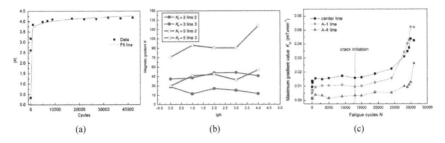

Fig. 3.3 Relationship between magnetic memory signals and fatigue cycles: **a** 18CrNi4A steel [20]; **b** 45CrNiMoVA steel [21]; and **c** 45 steel [22]

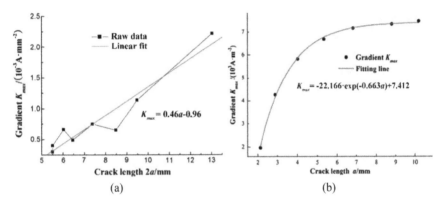

Fig. 3.4 Relationship between the magnetic memory signals and crack length: **a** Q345 structural steel [23]; and **b** Q235 weld joint [24]

the magnetic behavior was studied [25–27]. These research results indicated that MMM can detect the early stages of plastic deformation and effectively identify damage zones [28, 29].

- **Loading history**: The loading history and speed were also considered influencing variables in MMM testing by Zhejiang University in China [30, 31]. Experimental results indicated that the variation in the magnetic field not only depends on the existing damage state but also on the plastic deformation caused by the loading history [30].

- **Stress state**: Roskosz et al. [32, 33] from the Silesian University of Technology in Poland studied the distribution of residual stress using MMM. This method enabled the qualitative relationship between the residual magnetic field and residual stress to be determined.

It is worth noting that magnetic memory signals susceptible to numerous interference factors can also bring substantial difficulties for quantitative stress evaluation [34]. Li et al. [35] measured magnetic memory signals using installed magnetic sensor arrays and found that plastic deformation and residual stress around notches would increase the remnant flux leakage but that the effects were small. Gorkunov [36] also held that the affecting factors were numerous enough and that the contribution of each was rather difficult to differentiate and evaluate; thus, the reliability of evaluating the stress-strain state and applying MMM was low. In addition, Augustyniak et al. [37] constructed a magnetic finite element simulation and even proved that quantitative in situ NDT based on MMM was impossible. These results show that the reliability of MMM should be further improved from the perspective of engineering applications. Research concerning MMM is crucial as it provides obvious advantages in terms of early damage and service life assessment.

3.4 Standard Establishment

Due to the increasingly widespread application range of MMM in the welding field, the technical committee of the International Institute of Welding (IIW) published the ISO 24497 international standard of MMM in 2007 and revised the latest version in March 2020. This document was prepared by IIW, Commission V, NDT and Quality Assurance of Welded Products. This second edition cancels and replaces the first edition (ISO 24497-1:2007, ISO 24497-2:2007 and ISO 24497-3:2007), which has been technically revised and merged.

The new ISO 24497-1:2020 document specifies terms and definitions for NDT by the MMM technique as well as general requirements for the application of this technique of the magnetic testing method. The terms specified in this document are mandatory for application in all types of documentation and NDT literature using the MMM technique. MMM is defined as the magnetic state of a ferromagnetic object, depending on how the field has changed in the past and a consequence of the magneto-mechanical hysteresis of the material. It is worth noting that for a given magnetic field (e.g., the magnetic field of the earth), a ferromagnetic object formed in the course of its fabrication or in operation changes its residual magnetization due to diverse environmental factors that influence the magnetic domain distribution (e.g., temperature, mechanical loads or microstructural changes in the material). MMM testing is defined as the technique of the magnetic testing method in NDT based on the measurement and analysis of the magnetic stray field distribution on the surface of inspected objects without intentional (active) magnetization.

Based on this, MMM is closely related to MMM testing. The essence of MMM testing is a type of measurement and analysis of the magnetic stray field. This NDT technique can help to realize the following purposes:

(1) Determination of the heterogeneity of the magnetomechanical state of ferromagnetic objects and detection of defect concentrations and boundaries of metal microstructure heterogeneity; (2) determination of locations with magnetic stray field aberrations for further microstructural analysis and/or NDT and evaluation; (3) early diagnostics of fatigue damage of the inspected object and evaluation of its structural lifetime; (4) quick sorting of new and used inspection objects based on their magnetic heterogeneity for further testing; (5) efficiency improvement of NDT by

combining MMM testing with other NDT methods or techniques (ultrasonic testing, X-ray, etc.) by fast detection of the most likely defect locations; and (6) quality control of welded joints of various types and their embodiment (including contact and spot welding). See ISO 24497-2 for details of this application.

The new ISO 24497-2:2020 document specifies general requirements for the application of the MMM testing technique of the magnetic testing method for quality assurance of welded joints. This document can be applied to welded joints in any type of ferromagnetic product, such as pipelines, vessels, equipment, and metal construction, as agreed with the purchaser.

3.5 Applications for Remanufacturing

Besides, many experts have evaluated the damage degree of the remanufacturing core based on the achievements of experiments and theories combined with MMM. The relationship between the deformation of an axle housing and magnetic memory signals was established by Huang et al. [38], and fatigue crack propagation in a 510 L steel axle housing was also detected using MMM, as shown in Fig. 3.5, which provided solid references for remanufacturing [39]. Song et al. [40, 41] combined MMM with the X-ray diffraction technique and artificial neural network (ANN) to predict the plastic deformation rate and residual life of the axle housing of a waste drive. In addition, engine crankshafts also play a key role in remanufacturing. MMM can be used to monitor the locations of stress concentrations and fatigue crack propagation [42, 43], thereby providing a basis and guidance for the remanufacturing and life evaluation of retired crankshafts [44]. Fatigue cracking and stress concentrations easily appear on the inner face of a cylinder barrel with reciprocating motions of the

Fig. 3.5 Relationship between the stress intensity factor K_I, magnetic signal gradient K_{max} and number of fatigue cycles N [39]

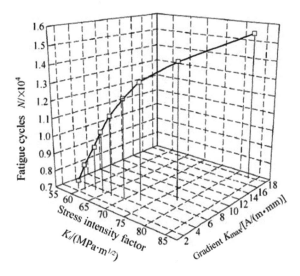

Fig. 3.6 Relationship
between the burial depth of a
crack, the load and the
magnetic signal gradient K
[50]

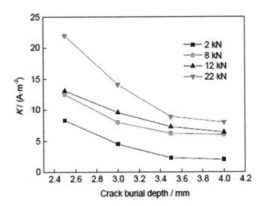

cylinder piston; thus, evaluating the quality of a cylinder barrel before remanufacturing is critical. Eddy current and magnetic memory tests were applied by Shi et al. [45, 46] to detect the superficial defects of a cylinder barrel. The results indicated that the composite detection technology can characterize cracks and stress and effectively guarantee the quality of a remanufactured cylinder. Moreover, many remanufactured parts, such as sucker rods [47], cylinder head bolts [48], turbine blades, gears and train wheels [49], have been detected using MMM.

When the remanufacturability of the remanufacturing core has been determined by NDT, it can be repaired by surface coating techniques; accordingly, repair quality evaluation is also an urgent problem. Liu et al. [50–53] evaluated the stress and crack propagation of laser cladding coatings via magnetic memory signals. Then, the relationship between the burial depth of a crack, the load and the normal component of the magnetic signal gradient value K was obtained [50], as shown in Fig. 3.6. Hua et al. [54] discussed the effects of fatigue damage caused by different laser cladding parameters and fatigue loads on magnetic memory signals. The results showed that fatigue damage location and degree can be predicted based on the changing curve of magnetic signals. Zeng et al. [55] proved that magnetic signals are closely related to the state of fatigue and thermal damage of laser cladding remanufactured products, which enables the quantitative assessment of the residual life of cladding layers to be carried out. In addition, Huang et al. [56] evaluated the cladding layers of plasma-transferred arc welding using MMM and discussed the variation in magnetic signals with stress.

3.6   Problems and Prospects

Recent demands for inspecting remanufacturing cores and increasing requirements for quality control in the repair process have led to the development of magnetic NDT techniques. Although these techniques, including the MMM method, have attracted considerable attention due to their unique advantages, the following problems and difficulties still exist with their current application in remanufacturing engineering:

- The quantitative criteria for core damage should be modified. The characteristic values of the testing signals can only qualitatively reflect the damage degree of the remanufacturing cores and approximately identify the location of defects. Consequently, accurately determining the remanufacturability of cores and the time when the damaged components should be remanufactured or repaired is challenging. Given these weaknesses, physical mechanisms and theoretical models based on the testing signals, as well as the signal characteristics, should be further clarified. Quantitative criteria have been key to establishing life prediction models for remanufacturing cores.
- Testing signals are easily subjected to interference. Variations in the environmental temperature and background magnetic field negatively affect the testing signals during detection and are responsible for the lack of accuracy and reliability of the detection results. The service conditions of remanufacturing cores are generally complex, thereby also resulting in high mistake and miss rates in the results. To improve the reliability of inspection, magnetic NDT techniques are expected to be combined to realize complementary advantages.
- The objects used for evaluation are not sufficiently comprehensive. Since the laser cladding technique is a popular research topic in the remanufacturing industry, repair quality evaluation of remanufactured products, mostly adopting MMM, has focused primarily on the laser cladding layers so far. The range of evaluation objects should also be extended to include coatings prepared by other surface repair technologies used in remanufacturing, including plasma spraying, plasma-transferred arc welding and electroplating deposition. In addition, more attention should be paid to cores with high remanufacturing value in nuclear, aerospace and military fields, which are likely to bring more economic benefits. Furthermore, since all of the remanufacturing forming processes directly relate to the bonding strength of strengthened coatings and their repair quality, it is also necessary to implement online monitoring in the processes.
- Engineering application should be reinforced. The objects of magnetic NDT are still restricted to specific parts or samples in laboratories, and there are few cases of magnetic NDT in remanufacturing engineering applications. Thus, the measurement criteria of magnetic NDT in remanufacturing must be established, a suitable detection system should be developed and integrated into remanufacturing, and personnel training must be strengthened.
- Relevant standards should be established and improved. International remanufacturing standards have not received much attention worldwide. For example, several examples of terminology in remanufacturing remain ambiguous or inconsistent, and various definitions of the term 'remanufacturing' are currently in circulation. Although the technical committee ISO/TC 135 has published over eighty international NDT standards, existing relevant magnetic NDT standards are not sufficiently sound, which impedes the application of magnetic NDT. Furthermore, the ISO 24497 standard of MMM was established by the International Institute of

Welding and may only be suitable for the evaluation of welded joints. Therefore, the standards of MMM have not yet been applied for remanufacturing and seldom refer to core damage assessment and repair quality evaluation. It is suggested that governments, enterprises and academics all over the world cooperate with one another to establish or improve a series of relevant international standards for magnetic NDT methods applied to remanufacturing.

Although there are few application cases of MMM in remanufacturing, this magnetic NDT technique has great potential value. Over the past few decades, many experts have made major contributions towards characterizing the material performance and defects by MMM. It is time to further combine MMM with remanufacturing and realize the core damage assessment and repair quality evaluation. Meanwhile, composite inspection technology in remanufacturing is recommended and should be improved in conjunction with other NDT techniques, such as Barkhausen noise, ultrasound, infrared and eddy current testing. It is predicted that key issues will be completely solved in the future with the development of composite magnetic NDT techniques.

References

1. A. Misra, Electromagnetic effects at metallic fracture. Nature **254**, 133–134 (1975)
2. A.A. Doubov, *Diagnostics of Boiler Tubes with Usage of Metal Magnetic Memory* (Energoatomizdat, Moscow, 1995)
3. D.C. Jiles, Theory of the magnetomechanicaleffect. J. Phys. D Appl. Phys. **28**(8), 1537–1546 (1995)
4. A.A. Doubov, Study of metal properties using magnetic memory method, in *Proceedings of the 7th European Conference on Nondestructive Testing*, Copenhagen (1998), pp. 920–927
5. J.L. Ren, J.M. Lin, *Metal Magnetic Memory Testing Technique* (China Electric Power Press, Beijing, 2000)
6. Z.D. Wang, Y. Gu, Y.S. Wang, A review of three magnetic NDT technologies. J. Magn. Magn. Mater. **324**(4), 382–388 (2012)
7. A.A. Doubov, Express method of quality control of a spot resistance welding with usage of metal magnetic memory. Weld. World **46**, 317–320 (2002)
8. A.A. Doubov, A technique for monitoring the bends of boiler and steam-line tubes using the magnetic memory of metal. Therm. Eng. **48**(4), 289–295 (2001)
9. L.Q. Zhong, L.M. Li, X. Chen, Simulation of magnetic field abnormalities caused by stress concentrations. IEEE Trans. Magn. **49**(3), 1128–1134 (2013)
10. W.S. Singh, R. Stegemann, M. Kreutzbruck, Three dimensional finite element analysis of the stress-induced geometry effect on self-magnetic leakage fields during tensile deformation. Insight: Non-Destruct. Test. Cond. Monit. **58**(10), 544–550 (2016)
11. P.P. Shi, K. Jin, X.J. Zheng, A general non-linear magnetomechanical model for ferromagnetic materials under a constant weak magnetic field. J. Appl. Phys. **119**(14), 196–203 (2016)
12. M.X. Xu, M.Q. Xu, J.W. Li et al., Using modified J-A model in MMM detection at elastic stress stage. Nondestruct. Test. Eval. **27**(2), 121–138 (2012)
13. Z.D. Wang, K. Yao, B. Deng et al., Theoretical studies of metal magnetic memory technique on magnetic flux leakage signals. NDT&E Int. **43**(4), 354–359 (2010)
14. Z.D. Wang, K. Yao, B. Deng et al., Quantitative study of metal magnetic memory signal versus local stress concentration. NDT&E Int. **43**(6), 513–518 (2010)

15. H.H. Huang, S.L. Jiang, C. Yang et al., Stress concentration impact on the magnetic memory signal of ferromagnetic structural steel. Nondestruct. Test. Eval. **29**(4), 377–390 (2014)

16. M.I.M. Ahmad, A. Arifin, S. Abdullah, Fatigue crack effect on magnetic flux leakage for A283 grade C steel. Steel Compos. Struct. **19**(6), 1549–1560 (2015)

17. P.P. Shi, X.J. Zheng, Magnetic charge model for 3D MMM signals. Nondestruct. Test. Eval. **31**(1), 45–60 (2016)

18. B. Liu, Y. Fu, B. Xu, Study on metal magnetic memory testing mechanism. Res. Nondestruct. Eval. **26**(1), 1–12 (2015)

19. B. Liu, Y. Fu, R. Jian, Modelling and analysis of magnetic memory testing method based on the density functional theory. Nondestruct. Test. Eval. **30**(1), 13–25 (2015)

20. C.L. Shi, S.Y. Dong, B.S. Xu et al., Stress concentration degree affects spontaneous magnetic signals of ferromagnetic steel under dynamic tension load. NDT&E Int. **43**(1), 8–12 (2010)

21. L.H. Dong, B.S. Xu, S.Y. Dong et al., Characterisation of stress concentration of ferromagnetic materials by metal magnetic memory testing. Nondestruct. Test. Eval. **25**(2), 145–151 (2010)

22. C.C. Li, L.H. Dong, H.D. Wang et al., Metal magnetic memory technique used to predict the fatigue crack propagation behaviour of 0.45%C steel. J. Magn. Magn. Mater. **405**, 150–157 (2016)

23. H.H. Huang, S.L. Jiang, R.J. Liu et al., Investigation of magnetic memory signals induced by dynamic bending load in fatigue crack propagation process of structural steel. J. Nondestrct. Eval. **33**(3), 407–412 (2014)

24. H.H. Huang, Z.C. Qian, C. Yang et al., Magnetic memory signals of ferromagnetic weldment induced by dynamic bending load. Nondestruct. Test. Eval. **32**(2), 166–184 (2017)

25. H.M. Li, Z.M. Chen, Y. Li, Characterisation of damageinduced magnetisation for 304 austenitic stainless steel. J. Appl. Phys. **110**(11), (2011)

26. H.M. Li, H.E. Chen, Z.S. Yuan et al., Comparisons of damage-induced magnetisations between austenitic stainless and carbon steel. Int. J. Appl. Electromagnet. Mech. **46**(4), 991–996 (2014)

27. J.W. Li, M.Q. Xu, Influence of uniaxial plastic deformation on surface magnetic field in steel. Meccanica **47**(1), 135–139 (2012)

28. J.C. Leng, Y. Liu, G.Q. Zhou et al., Metal magnetic memory signal response to plastic deformation of low carbon steel. NDT&E Int. **55**(3), 42–46 (2013)

29. J.C. Leng, M.Q. Xu, M.X. Xu et al., Magnetic field variation induced by cyclic bending stress. NDT&E Int. **42**(5), 410–414 (2009)

30. S. Bao, S.N. Hu, L. Lin et al., Experiment on the relationship between the magnetic field variation and tensile stress considering the loading history in U75V rail steel. Insight: Non-Destruct. Test. Cond. Monit. **57**(12), 683–688 (2015)

31. S. Bao, Y.B. Gu, M.L. Fu et al., Effect of loading speed on the stress-induced magnetic behaviour of ferromagnetic steel. J. Magn. Magn. Mater. **423**, 191–196 (2017)

32. M. Roskosz, P. Gawrilenko, Analysis of changes in residual magnetic field in loaded notched samples. NDT&E Int. **41**(7), 570–576 (2008)

33. M. Roskosz, M. Bieniek, Evaluation of residual stress in ferromagnetic steels based on residual magnetic field measurements. NDT&E Int. **45**(1), 55–62 (2012)

34. W.M. Zhang, Z.C. Qiu, J.J. Yuan et al., Discussion on stress quantitative evaluation using metal magnetic memory method. J. Mech. Eng. **51**(8), 9–13 (2015)

35. Z.C. Li, S. Dixon, P. Cawley et al., Study of metal magnetic memory (MMM) technique using permanently installed magnetic sensor arrays. AIP Conf. Proc. **1806**(1), (2017)

36. E.S. Gorkunov, Different remanence states and their resistance to external effects. Discussing the so-called magnetic memory method. Insight: Non-Destruct. Test. Cond. Monit. **57**(12), 709–717 (2015)

37. M. Augustyniak, Z. Usarek, Discussion of derivability of local residual stress level from magnetic stray field measurement. J. Nondestr. Eval. **34**(3), 21 (2015)

38. H.H. Huang, J.Y. Yao, R.J. Liu et al., Damage detection of axle housing based on metal magnetic memory testing technology. J. Electron. Meas. Instrum. **28**(7), 770–776 (2014)

39. H.H. Huang, R.J. Liu, X. Zhang et al., Magnetic memory testing towards fatigue crack propagation of 510L steel. J. Mech. Eng. **49**(1), 135–141 (2013)

40. S.X. Song, J.R. Zhao, Z.Y. Tang et al., Plastic damage research for drive axle housing remanufacturing. China Mech. Eng. **24**(4), 538–541 (2013)

41. S.X. Song, J.R. Zhao, T. Liu, The prediction for the residual life of waste drive axle housing basing on neural network. Adv. Mater. Res. **308–310**, 246–250 (2011)

42. N. Xue, L.H. Dong, B.S. Xu et al., Characterisation of fatigue damage of crankshaft remanufacturing core by two dimensional magnetic memory signal. Adv. Mater. Res. **538–541**, 1588–1593 (2012)

43. J.B. Liao, H.F. Zhou, D. Sun, Research for the crack dynamic monitoring of crank shaft of marine diesel engine based on magnetic memory technology. Inf. Technol. J. **11**(4), 516–519 (2012)

44. C. Ni, L. Hua, X.K. Wang et al., Coupling method of magnetic memory and eddy current non-destructive testing for retired crankshafts. J. Mech. Sci. Technol. **30**(7), 3097–3104 (2016)

45. C.L. Shi, S.Y. Dong, B.S. Xu et al., Eddy current and metal magnetic memory testing for superficial defects of old cylinder barrel before remanufacturing, in *Proceedings of the 9th National Annual Conference on Nondestructive Testing* (2010), pp. 624–629

46. C.L. Shi, S.Y. Dong, W.X. Tang et al., Eddy current and metal magnetic memory testing nondestructive evaluation for superficial defects of old cylinder barrel. Mater. Sci. Forum **850**, 107–112 (2016)

47. J.C. Leng, Z.M. Wu, H. Zhang et al., Research progress of detection on remanufacturability of sucker rod. Mater. Rev. **30**(8), 62–67 (2016)

48. J.H. Yu, Magnetic memory testing and intensity failure analysis of fractured bolt of cylinder head on hydrogen gas compressor. Adv. Mater. Res. **503–504**, 772–775 (2012)

49. Y.L. Zhang, D. Zhou, P.S. Jiang et al., The state-of the-art surveys for application of metal magnetic memory testing in remanufacturing. Adv. Mater. Res. **301–303**, 366–372 (2011)

50. B. Liu, K. Gong, Y.X. Qiao et al., Evaluation of influence of present crack burial depth on stress of laser cladding coating with metal magnetic memory. Acta Metall. Sin. **52**(2), 241–248 (2016)

51. B. Liu, S.J. Chen, S.Y. Dong et al., Stress measurement of laser cladded ferromagnetic coating with metal magnetic memory. Trans. China Weld. Instit. **36**(8), 23–26 (2015)

52. B. Liu, Z.J. Shao, S.Y. Dong, Research on depth evaluation of surface crack in laser cladding coating with metal magnetic memory technology. Appl. Mech. Mater. **651–653**, 73–77 (2014)

53. B. Liu, *Ultrasonic and Metal Magnetic Memory Testing Method for Stress and Flaw Non-Destructive Evaluation of Remanufacturing Metal Coating* (Harbin Institute of Technology, Harbin, China, 2012)

54. L. Hua, W. Tian, W.H. Liao et al., Experimental study of laser cladding specimen under fatigue load by magnetic memory methods. Mach. Des. Manuf. **8**, 186–189 (2014)

55. C. Zeng, W. Tian, L. Hua, Characterisation of metal magnetic memory of thermal damage correlated with initial cumulative fatigue damage for laser cladding remanufacturing technology. Key Eng. Mater. **525–526**, 77–80 (2012)

56. H.H. Huang, G. Han, C. Yang et al., Stress evaluation of plasma-sprayed cladding layer based on metal magnetic memory testing technology. J. Mech. Eng. **52**(20), 16–22 (2016)

Part II
Detection of Damage in Ferromagnetic Remanufacturing Cores by the MMM Technique

Chapter 4
Stress Induces MMM Signals

4.1 Introduction

A large number of components, such as engine crankshafts, high-speed bearings and gears, are usually subjected to complex loads in service. Naturally, stress can be induced in these components by the applied tensile, compressive and bending loads. In addition, the stress level increases with the increase in loads or fatigue cycles. When the stress exceeds a certain level, the crack will initiate and propagate at the stress concentration zone, which can affect the performance and safety of the components. Therefore, it is necessary to evaluate the stress level to determine the damage degree or residual life, which can help us determine whether the damaged components need to be remanufactured.

The metal magnetic memory (MMM) testing method has attracted much attention because of its potential for microcrack and early stress damage evaluation. Based on the introduction in Chaps. 1–3, we know that the MMM method utilizes the stress-induced surface spontaneous magnetic flux leakage of ferromagnetic materials under the excitation of a geomagnetic field to investigate its stress distribution. The variations in MMM signals have been studied by many researchers based on different kinds of characteristic values, including the normal component $H_p(y)$, tangential component $H_p(x)$, gradient K, etc. [1–6].

The key point of quantization for the damage degree of remanufacturing core is establishing the relationship between the magnetic signal characteristics and stress levels. Shi et al. [7] proposed an inverse model including the objective function and optimization parameters to establish a reconstruction method, which can be used to invert stress and cracking in ferromagnetic materials. However, this method does not provide a clear explanation of the mechanism of MMM. Besides, previous studies on the MMM method focused on the variation in the MMM signal under static tensile stresses, and the magnetomechanical effect and the physical mechanism of the MMM phenomenon were also discussed in static tensile experiments. Generally, when the static stress exceeds the yield strength of materials, the main damage form of remanufacturing cores is the plastic deformation. When the static stress reaches the

© Science Press 2021

H. Huang et al., *Metal Magnetic Memory Technique and Its Applications in Remanufacturing*, https://doi.org/10.1007/978-981-16-1590-0_4

ultimate strength, the crack occurs immediately which leads to the fracture. However, the damage of remanufacturing cores accumulates slowly and the crack undergoes a long propagation process under the effect of fatigue stress. As a result, the variation in magnetization under cyclic stress is different from that under static stress [8, 9]. It is necessary to distinguish the variations in MMM signals under static stress and cyclic stress. And it is also important to determine how this MMM technique is used to detect deformation in tensile loads and crack propagation in the dynamic fatigue process of ferromagnetic materials.

This chapter was prepared to explore the variations in the MMM signals of ferromagnetic components induced by static tensile stress and cyclic bending stress, as shown in Sects. 4.2 and 4.3, respectively. The variations in the MMM signals were analyzed, the accumulation processes of the stress at different loads and numbers of fatigue cycles were investigated, and the physical mechanism of magnetization induced by stress was discussed.

4.2   Variations in the MMM Signals Induced by Static Stress

In Ref. [10], Bao et al. selected the samples made of Q235 low-carbon steel (corresponding to ASTM A283-C) to investigate the variations in the MMM signals under static stress in the elastic stage and plastic stage. The samples were demagnetized in the initial state. The chemical composition of the steel is listed in Table 4.1. The yield strength of the steel is approximately 220–230 MPa, and the ultimate strength is approximately 375–460 MPa. The dimensions of the specimen with a thickness of 6 mm are shown in Fig. 4.1. The tensile force was applied parallel to the longitudinal axis of the sample at a strain rate of 0.2 mm/min at room temperature. A steel sample was loaded to a predetermined stress level before the test was interrupted, and the sample was dismantled from the testing machine. The MMM signals were

Table 4.1 Chemical composition (wt%) of Q235

Material	C	Mn	Si	P	S
Q235	0.14–0.22	≤1.4	≤0.35	≤0.045	≤0.050

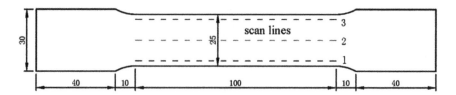

Fig. 4.1 Shape and dimension of a test specimen (in mm) [10]

then recorded along the three scanning lines shown in Fig. 4.1. As the measurements along the three scanning lines were almost the same, only the results of line 2 were discussed in this research. The sample was reinstalled onto the testing machine and loaded to the next higher stress level after the magnetic measurement was taken. This testing procedure was repeated until the sample fractured.

4.2.1 Under the Elastic Stage

Figure 4.2 shows the variations in the normal component $H_p(y)$ and the tangential component $H_p(x)$ of the sample loaded from 0 kN to 30 kN in the elastic stage. From Fig. 4.2a, one may observe that the fluctuation of the magnetic field $\Delta M1$ is approximately 150 A/m at 0 kN. The magnetic fields of the initial load-free state may be associated with the process of machining. As the applied load increases, the magnetic fluctuation tends to decrease. The magnetic fluctuation of $\Delta M2$ is reduced to approximately 30 A/m when the load increases to 30 kN. The same phenomenon can also be observed in the $H_p(x)$ component in Fig. 4.2b. The initial trace of $H_p(x)$ at 0 kN resembles a sinusoidal waveform. The amplitude of waveform $\Delta M3$ is reduced from 95 A/m at 0 kN to $\Delta M4$ of 20 A/m at 30 kN. Figure 4.2 demonstrates that the fluctuation of the magnetic fields, ΔM, along the length of the sample decreases as the applied load increases in the elastic stage. Figure 4.3 illustrates the variation in the magnetic domains with increasing load. In the initial state, the orientations of the magnetic domains are disordered, as shown in Fig. 4.3a. As the applied load increases, the residual stress and imperfection in the sample begin to adjust, and the irregular magnetic domains tend to rotate towards the easy magnetization direction, gradually turning to the same ordering, as shown in Fig. 4.3b, c. This leads to the distribution of the magnetic field being almost linear and varying much less than that in the initial stage.

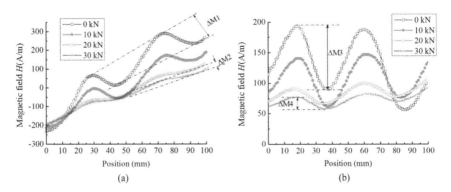

Fig. 4.2 MMM signals of the sample at various applied loads in the elastic stage: **a** normal component $H_p(y)$ and **b** tangential component $H_p(x)$ [10]

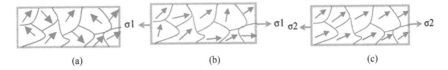

Fig. 4.3 Orientation of the magnetic domains: **a** initial orientation of the domains; **b** orientation of the domains with increasing stress; and **c** domains turning to the same ordering with a further increase in stress [10]

4.2.2   Under the Plastic Stage

When the applied load is above 40 kN, the sample is in the plastic stage. When the sample is reloaded to 66 kN, it starts to neck and is elongated from 100 to 130 mm. Therefore, the variations in the normal component $H_p(y)$ and the tangential component $H_p(x)$ of the sample loaded from 40 kN to 66 kN in the plastic stage are shown in Fig. 4.4. In Fig. 4.4a, one may observe that the fluctuation of the magnetic field further decreases as the applied load increases, and with a tensile load of 66 kN, the magnetic trace becomes approximately a straight line with a magnetic gradient of 3 A/m/mm. The magnetic curves at various stress levels are almost parallel to each other. In Fig. 4.4b, the magnetic curves demonstrate similar behavior to that shown in Fig. 4.4a. The magnetic trace of $H_p(x)$ finally evolves into an almost straight line with a magnetic gradient of 0.5 A/m/mm at a tensile load of 66 kN. From Figs. 4.2 and 4.4, one may observe that the magnetic fields change dramatically with increasing applied loads in the elastic stage but remain relatively stable in the plastic stage. This may be associated with the dislocation activity that occurs during the loading process. The magnetic properties of the steel specimen are closely related to the applied stress as well as to the stress-induced microstructural defects formed in the material. In the elastic loading stage, few microstructural defects initiate in

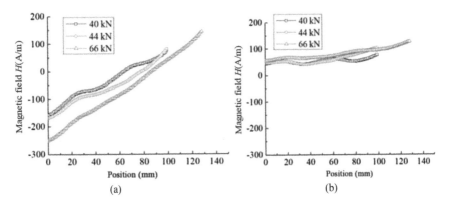

Fig. 4.4 MMM signals of the sample at various applied loads in the plastic stage: **a** normal component $H_p(y)$ and **b** tangential component $H_p(x)$ [10]

the material to impede magnetic domain wall movement. The applied stress induces magnetic domain reorientation so that the stress-induced magnetoelastic energy of the magnetic domain is minimized. As a result, the bulk magnetic fields demonstrate drastic variations as the external stress increases. In the plastic loading stage, however, the magnetic field demonstrates different behavior compared to that in the elastic stage.

When the sample was finally reinstalled on the testing machine and reloaded, it ruptured at 64 kN. Figure 4.5a shows the scanning electron microscope (SEM) micrograph of the fracture surface of the sample. One may observe that there are many dimples with different shapes and sizes, and there are second phase particles in these dimples. This indicates that the tensile test produces ductile failure. To explore the relationship between the stress-induced magnetic field and the dislocations occurring during the loading process, a transmission electron microscope (TEM) study of the sample at the fracture section was carried out, and the result is shown in Fig. 4.5b. It can be observed that the dislocation configuration in the grain demonstrates many uneven regions with high dislocation density or even with dislocation tangles, i.e., dark bands. Figure 4.5a, b indicate that the Q235 steel specimen showed large plastic deformation and active dislocation during the plastic loading stage. Both the applied stress and the microstructural defects induced by plastic deformation affect the magnetic properties of the specimen. The increased dislocation density pinned the magnetic domain and impeded the magnetic domain ordering process. The hindering of domain wall movement caused by dislocations is much stronger than the reorientation of the magnetic domain induced by the external stress. As a result, the measured magnetic fields in the plastic stage remained relatively stable.

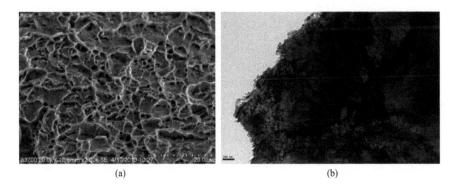

(a)   (b)

Fig. 4.5 Micrograph of the fracture surface of the sample: **a** SEM and **b** TEM [10]

4.2.3   Theoretical Analysis

The different change laws of MMM signals under elastic and plastic stages can be inferred by the magnetomechanical model introduced in Chap. 2. Based on the research of Li et al. [11], the effective field can be calculated and plotted in Fig. 4.6a. The effect of elastic stress on the ferromagnetic material is equivalent to a magnetic field. The overall amplitude of the magnetic field near the sample surface is larger at the first stage and then decreases as a response to increasing tensile stress. The modeling results qualitatively confirm the experimental data in Fig. 4.2. The variation in the magnetization will always approach the anhysteretic magnetization case. The magnitude of the magnetization change at the preliminary stage of loading was larger than that at the late stage, as shown in Fig. 4.6b. Therefore, the magnetic fluctuation in the experiment also tends to decrease as the applied load increases.

Based on the magnetoplastic model, the total effective field can be expressed as:

$$H_{eff} = H + \alpha M - D_\sigma M + H_{\sigma r} + H_{\sigma p} \tag{4.1}$$

where H is the applied magnetic field, α is the mean field coefficient representing the self-coupling of the magnetization, $D_\sigma M$ is the stress demagnetization term, $H_{\sigma r}$ is the contribution of residual stress and $H_{\sigma p}$ is the effect of the plastic deformation on the effective field. The hysteresis behavior of polycrystalline materials can be derived by using a model of magnetic moments interacting with the effective field. Hence, the magnetization of a bulk magnetic body can be calculated from the following equation:

$$\frac{M}{M_s} = \coth\left(\frac{H_{eff}}{a}\right) - \frac{a}{H_{eff}} + \frac{k\delta}{a}\left\{\cos \mathrm{ech}^2\left(\frac{H_{eff}}{a}\right) - \left(\frac{a}{H_{eff}}\right)^2\right\} - k\delta\left(\frac{\partial^2 M}{\partial^2 H_{eff}}\right)$$

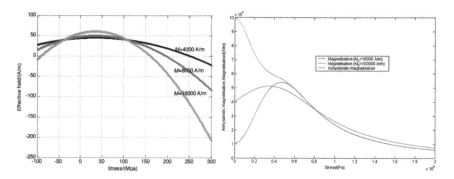

Fig. 4.6 Theoretical calculation based on the magnetomechanical model: **a** variation in the effective field with static stress; **b** variation in magnetization with static stress [11]

Fig. 4.7 Curves of the magnetization versus plastic strain under an applied magnetic field of 40 A/m [11]

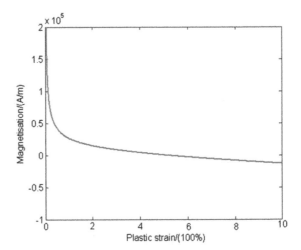

$$\approx \coth\left(\frac{H_{eff}}{a}\right) - \frac{a}{H_{eff}} + \frac{k\delta}{a}\left\{\cos\text{ech}^2\left(\frac{H_{eff}}{a}\right) - \left(\frac{a}{H_{eff}}\right)^2\right\}$$

$$+ \frac{2k^2}{a^2}\left\{\cos\text{ech}^2\left(\frac{H_{eff}}{a}\right)\coth\left(\frac{H_{eff}}{a}\right) - \left(\frac{a}{H_{eff}}\right)^3\right\} \quad (4.2)$$

where a is a scaling constant, k is the pinning coefficient and $\delta = \pm 1$ for increasing and decreasing magnetic fields, respectively. Figure 4.7 shows the influence of the plastic deformation on the magnetization under an applied magnetic field of 40 A/m. A larger change appears at a small plastic deformation, after which the values of the magnetic field decrease slowly with increasing plastic strain. The theoretical results are also in good agreement with the experimental observations in the plastic stage, as shown in Fig. 4.4.

4.3 Variations in the MMM Signals Induced by Cyclic Stress

For the research of MMM signals under cyclic stress, a sample of Q345 low-carbon steel (corresponding to ASTM Gr·50) is selected due to its excellent plasticity and welding performance, which is widely used in automobiles, pressure vessels, bridges and hoisting machinery. Its yielding strength is 358 MPa, its ultimate strength is 484 MPa, and its chemical constitution is shown in Table 4.2. Eight specimens with

Table 4.2 Chemical composition (wt%) of Q345

Material	C	Mn	Si	P	S	Al	V	Nb	Ti
Q345	≤0.2	1.0–1.6	≤0.55	≤0.035	≤0.035	≥0.015	0.02–0.15	0.015–0.06	0.02–0.2

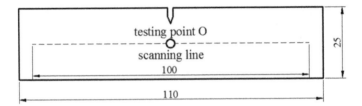

Fig. 4.8 Shape of the sheet specimen (in mm), scanning line, and X-ray testing point

a center notch length of 5.5 mm were fabricated, as shown in Fig. 4.8. The scanning line with a length of 100 mm was marked on each specimen.

Three-point bending fatigue tests were carried out on an MTS809 hydraulic servo machine, whose dynamic load error was within ±1.0%. Dynamic bending loading was performed with a constant amplitude (sinusoidal waveform), with the maximum bending moments at 151 Nm, the minimum bending moments at 15 Nm, and frequency at 10 Hz. The magnetic probe with a 1A/m sensitivity based on the Hall sensor was installed on a nonferromagnetic 3D electric scanning platform and was placed vertical to the surface of the specimen with a lift-off value of 1 mm. When loaded to the preset cycle number, the specimen was taken from the holders and laid on the platform in the south to north direction, and the normal component $H_p(y)$ of the magnetic signal values at the scanning line was measured. X-ray diffraction detection was conducted at point O of the specimens, as shown in Fig. 4.8. The surface residual stress and the diffraction peak width of each specimen were collected. Pearson VII distribution functions were adopted to determine the X-ray diffraction peak width, and the X-ray penetration depth was 3–10 μm. The length of the fatigue cracks in the specimen was measured by a JXD-250B reading microscope. Fatigue crack propagation of the specimens was investigated by a JSM-6700F scanning electronic microscope (SEM).

4.3.1 Under Different Stress Cycle Numbers

Figure 4.9 shows the variable curves of the magnetic signals $H_p(y)$ from the scanning line of the specimens experiencing different fatigue load cycles. The initial curve from specimen #1 presented good linearity, and the $H_p(y)$ value decreased from 30 to −45 A/m, which came from the residual magnetic field during manufacturing. To eliminate the impact of the initial magnetic field, the specimens were demagnetized before loading.

The shapes of the MMM signal $H_p(y)$ curves remained relatively unchanged during the fatigue tests, as shown in Fig. 4.9. Jiles developed a model to describe the stress dependence of magnetization, which shows that the rate of change of magnetization not only depends on stress but also anhysteretic magnetization [12]. Leng et al. suggested that the surface magnetic fields induced by stress contain reversible

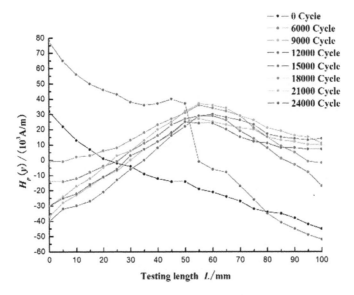

Fig. 4.9 Magnetic signals from scanning lines for eight specimens experiencing different fatigue load cycles: (1) specimen #1 with no loads; (2) specimen #2 at 6000 cycles; (3) specimen #3 at 9000 cycles; (4) specimen #4 at 12,000 cycles; (5) specimen #5 at 15,000 cycles; (6) specimen #6 at 18,000 cycles; (7) specimen #7 at 21,000 cycles; and (8) specimen #8 at 24,000 cycles when the specimen was about to fracture and fail

and irreversible processes prior to failure in a rotary bending fatigue experiment of 45-steel [8]. The magnetization is reduced with applied cyclic stress to overcome the internal friction forces so that it could approach its anhysteretic state, which is an irreversible process. Once most weak domain wall pinning sites were overcome, the magnetization of the specimen reached anhysteretic magnetization. However, this is mostly a reversible process because domain walls remaining on strong pinning sites do not become unpinned under the effect of cyclic stresses [8, 13]. The study on the static tensile test in Sect. 4.2 also confirmed that the $H_p(y)$ curve changed rapidly at the beginning and tended to a relatively stable stage after the stress was greater than the yield strength, which was in the range of 160–200 MPa [14, 15]. These findings were consistent with the test results.

When the applied loading cycles reached 24,000 cycles, the specimen was about to fracture, and the magnetic signals intensively changed polarity and had zero value in the scanning line. This phenomenon was explained by Shi [7] and Leng [8] in both cyclic tensile tests and cyclic bending tests. In terms of the interaction energy in a ferromagnetic material, the total free energy includes magnetocrystalline anisotropy energy E_k, stress energy E_σ and demagnetization energy E_d when the test temperature is far below the Curie point, provided that the total free energy E is equal to the internal energy, i.e., $E = E_k + E_\sigma + E_d$. For isotropic magnetostrictive materials, the stress energy E_σ can be given as follows [16]

$$E_\sigma = -3/2\lambda_s \sigma \cos^2\theta \qquad\qquad (4.3)$$

where σ is stress, λ_s is magnetostriction, and θ is the angle between the stress axis and the direction of magnetization. E_σ is far greater than E_k and E_d under the applied axial stress. When the specimen was about to fracture, the stress energy was released immediately. At the same time, the demagnetization energy increased dramatically to recover the system to a balanced state in terms of thermodynamic equilibrium. Therefore, positive-negative magnetic poles occurred on the fracture zone, and the amplitude of the MMM signals changed sharply.

It should be noted that the maximum values of $H_p(y)$ always appeared at distances of 50–55 mm where the notches were fabricated before the specimens fractured, which was different from the results reported for the tensile tests in Sect. 4.2. The variation in $H_p(y)$ illustrated the behavior of the magnetic field of the surface above the neutral plane, which was under compressive stress in the dynamic bending tests. The magnetization under tensile stress is different from that under compressive stress, as shown in Fig. 4.10. It is known that applied mechanical stress can result in changes in the domain structure and therefore the internal magnetic field of ferromagnetic materials due to the magnetomechanical effect [4, 17], which induces residual magnetic signals. Figure 4.10 shows the domain's reorientation influenced by tensile stress and compressive stress. When the specimen is loaded under the geomagnetic field, the domains turn along the axial direction of the applied tensile stress or perpendicular direction of the applied compressive stress based on the piezomagnetic effect. This includes movement of the domain walls, increasing the domains with favored magnetic direction and annihilating the others, and the domains' rotation. Furthermore, the ferromagnetic clamp of the testing machine may have an impact on the distribution of the magnetic signals throughout the three-point bending tests so that the maximum values of $H_p(y)$ did not appear exactly on the notch position.

To describe the variation in the MMM signal at the scanning line, the gradient of the $H_p(y)$ curves, K, was given as follows:

$$K = \left|\Delta H_p(y)/\Delta L\right| \qquad\qquad (4.4)$$

where $\Delta H_p(y)$ is the differential value of the magnetic signals between two points and ΔL is the distance between the two points. K_{max} is the maximum gradient of $H_p(y)$ at the scanning line of each specimen. K_{max} always appeared in the middle of the specimen in the tests, as shown in Fig. 4.11a, as the magnetic signals changed

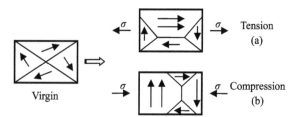

Fig. 4.10 Stress affects the movement of magnetic domains in the microstructure with microdefects under the geomagnetic field: **a** under tensile stress and **b** under compressive stress

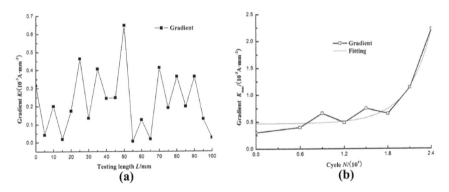

Fig. 4.11 Gradient of the magnetic signal $H_p(y)$: **a** gradient K at the scanning line at 18,000 loading cycles and **b** maximum gradient K_{max} at different cycles

intensively at the location of the notch. Figure 4.11b presents the values of K_{max} at different cycles for 8 specimens and their fit line. It can be seen that K_{max} is similar to the exponential variation. The relation between K_{max} and loading cycles can be fitted by the following exponential fitting equation.

$$K_{max} = a_0 \times exp(N/t) + k_0 \tag{4.5}$$

where N is the loading cycle and a_0, t and k_0 are constant values related to the loading type, loading value and material. In the experiment, a_0 is 0.00116, t is 0.32797, and k_0 is 0.46259; the residual standard deviation (measurement versus fitting) is 0.408, and R-squared (coefficient of determination) is 0.961.

4.3.2 Characterization of Fatigue Crack Propagation

As the number of fatigue loading cycles increased, the stress accumulated continuously, which led to the three stages of crack propagation, as shown in Fig. 4.12. Cracks initiated at the tip of the notch under a dynamic bending load. In the early

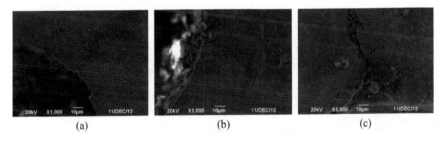

Fig. 4.12 Fatigue crack propagation at the crack tip at three stages: **a** in the early stage at a loading cycle of 6000; **b** in the intermediate stage at a loading cycle of 9000; and **c** in the late stage at a loading cycle of 21,000

stage of crack propagation, the stress accumulation remained at a lower level, and fatigue cracks were only observed by using the JXD-250B reading microscope at a loading cycle of 6000. In the intermediate stage, with the increase in the number of loading cycles, the stress level also increased, the crack propagated to some length at loading cycles from 9000 to 18,000, and crack branching appeared, as shown in Fig. 4.12b. In the late stage, the stress accumulated further, and the specimen was about to fail at loading cycles from 21,000 to 24,000.

The stress intensity factor, K_I, which has a linear relationship with the load and depends on the geometric shape and size of the specimen, is a critical parameter for describing the stress and strain field at the crack tip [18]. K_I of the specimen in this experiment can be expressed as follows.

$$K_I = \frac{6M}{\sqrt{(h - 2a)^3}} g(2a/h) \tag{4.6}$$

where $2a$ is the length of the crack, h is the width of the specimen, M is the bending moment, and $g(2a/h)$ is the shape factor.

The relationships between $2a$, K_I, and K_{max} at different cycles are shown in Fig. 4.13a, b. It shows that $2a$, K_I and K_{max} had the same variation trend increasing with the increase in the number of load cycles. In the early and intermediate stages of crack propagation, the crack propagated slowly with the increase in the number of load cycles, and K_{max} fluctuated in the interval of 0.3–0.75 × 10⁻³ A/m². In the late stage, the crack rapidly propagated to 13.023 mm, which was 52.1% of the specimen width, and K_{max} increased intensively, reaching 2.22 × 10⁻³ A/m² at 24,000 cycles. When fatigue cracks developed, more magnetic charges accumulated on the two sides of the fatigue cracks due to the reorientation of more domains, and the MMM signals varied intensively to output a higher K_{max} value; therefore, a similar regularity of K_{max} and K_I was confirmed in the crack propagation process.

According to the experimental results, it is believed that K_{max} was related to the length of the fatigue crack. Figure 4.13c shows the analysis results. It can be seen that K_{max} was approximately linear to crack length $2a$, which can be described by:

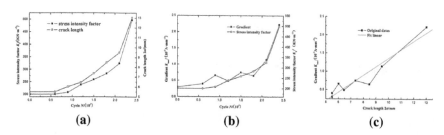

(a)   (b)   (c)

Fig. 4.13 Relationship between **a** the stress intensity factor and crack length; **b** the stress intensity factor and maximum gradient of the magnetic signals; and **c** the maximum gradient and crack length after fitting a straight line

$$K_{max} = k \times a + b \qquad (4.7)$$

where k and b are both constant values related to the material and shape of the specimen. In the experiment, k is 0.46, and b is -0.96. It can be seen that K_{max} and $2a$ were accurately fitted by Eq. (4.7).

There are many factors impacting the K_{max} value, such as the load (load type, load value and load position), specimens (material, geometry and dimension), and manufacturing process of the specimens (welding, casting, forging, heat treatment, etc.). The above formula is still an approximation expression, and more experimental work is needed to modify it. However, the results indicate that it is feasible to determine the fatigue crack propagation for structural steel under three-point bending load conditions.

Furthermore, the plastic deformation in the crack tip, which can reflect the residual stress level, can also be investigated by the X-ray diffraction method with the X-ray diffraction peak width [19]. Figure 4.14 shows that the X-ray diffraction peak width and the maximum gradient K_{max} had the same variation trend. It has been established that the average grain or subgrain volume decreases due to plastic deformation, and consequently, the number of subgrains increases. Since microscopic stress and an increasing number of subgrains occurred simultaneously in the crack propagation process of the specimen, stress induced the MMM signal on the surface of the specimen, and a change in crystallite size led to a variation in the width of the X-ray diffraction peak. Both the range of the diffraction peak width variation and K_{max} value were small in the early stage of crack propagation when plastic deformation started occurring in the incision area under dynamic bending load. There was a large jump in both the diffraction peak width and K_{max} in the late stage when the specimen was about to fracture. Therefore, MMM/X-ray diffraction combined detection can be beneficial for confirming the fatigue stress level and crack propagation process of a ferromagnetic component under a dynamic bending load.

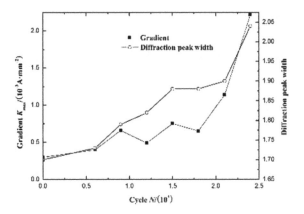

Fig. 4.14 Relationship between K_{max} and the X-ray diffraction peak width detected at testing point O

4.4   Conclusions

In this chapter, the variations in the MMM signals in ferromagnetic materials induced by static and cyclic stresses are investigated. The static stress-induced magnetic fields demonstrate different characteristics in the elastic and plastic loading stages. In the elastic regime, the amplitude of the magnetic field variation decreases with increasing external stress due to magnetic domain reorientation that minimizes the magnetoelastic energy. In the plastic regime, the magnetic field pattern remains relatively stable with increasing stress, which can be attributed to the fact that the pinning effect of dislocations in this stage is much stronger than the domain reorientation effect due to external stress. However, during the fatigue process, the MMM signals were relatively stable at different loading cycles, and the maximum value appeared at the fatigue crack area before the specimen fractured. The magnetic signal maximum gradient value K_{max}, which can be used to identify the degree of the stress level, was found to vary exponentially with increasing number of loading cycles and was approximately linear with respect to the crack length. K_{max} is potentially a very useful indicator for monitoring the fatigue stress level and crack propagation for ferromagnetic structural steel under bending loads. However, quantifying the K_{max} value is still a new problem, and there are many issues regarding the detection mechanism that need to be addressed further. The feasibility of the combined MMM/X-ray diffraction testing method for detecting fatigue stress damage is our future work.

References

1. L.H. Dong, B.S. Xu, S.Y. Dong et al., Variation of stress induced magnetic signals during tensile testing of ferromagnetic steels. NDT&E Int. **41**, 184–189 (2008)
2. L.H. Dong, B.S. Xu, S.Y. Dong et al., Stress dependence of the spontaneous stray field signals of ferromagnetic steel. NDT&E Int. **42**, 323–327 (2009)
3. J.C. Leng, Y. Liu, G.Q. Zhou et al., Metal magnetic memory signal response to plastic deformation of low carbon steel. NDT&E Int. **55**, 42–46 (2013)
4. C.L. Shi, S.Y. Dong, B.S. Xu et al., Metal magnetic memory effect caused by static tension load in a case-hardened steel. J. Magn. Magn. Mater. **322**, 413–416 (2010)
5. Z.D. Wang, K. Yao, B. Deng et al., Quantitative study of metal magnetic memory signal versus local stress concentration. NDT&E Int. **43**, 513–518 (2010)
6. M. Roskosz, P. Gawrilenko, Analysis of changes in residual magnetic field in loaded notched samples. NDT&E Int. **41**, 570–576 (2008)
7. P.P. Shi, K. Jin, P. Zhang et al., Quantitative inversion of stress and crack in ferromagnetic materials based on metal magnetic memory method. IEEE Trans. Magn. **54**, 1–11 (2018)
8. J.C. Leng, M.Q. Xu, M.X. Xu et al., Magnetic field variation induced by cyclic bending stress. NDT&E Int. **42**, 410–414 (2009)
9. J.W. Li, M.Q. Xu, M.X. Xu, Investigation of the variation in surface magnetic field induced by cyclic tensile-compressive stress. Nondestruct. Test. Eval. **27**, 1–7 (2012)
10. S. Bao, S. Gong, Magnetic field variation of a low-carbon steel under tensile stress. Insight **56**, 252–255 (2014)
11. J.W. Li, L.Y. Xiao, X.M. Han, Influence of applied load in ferromagnetic material on magneto-mechanical phenomena. Insight **58**, 308–312 (2016)

12. D.C. Jiles, Theory of the magnetomechanical effect. J. Phys. D Appl. Phys. **28**, 1537–1546 (1995)
13. Y. Chen, B.K. Kriegermeier-Sutton, J.E. Snyder et al., Magnetomechanical effects under torsional strain in iron, cobalt and nickel. J. Magn. Magn. Mater. **236**, 131–138 (2001)
14. E. Yang, L.M. Li, X. Chen, Magnetic field aberration induced by cycle stress. J. Magn. Magn. Mater. **312**, 72–77 (2007)
15. P.J. Guo, X.D. Chen, W.H. Guan et al., Effect of tensile stress on the variation of magnetic field of low-alloy steel. J. Magn. Magn. Mater. **323**, 2474–2477 (2011)
16. T. Miyazaki, H.M. Jin, *The Physics of Ferromagnetism* (Springer, Berlin Heidelberg, 2012)
17. W.Y. Gong, Z.M. Wu, H. Lin et al., Longitudinally driven giant magneto-impedance effect enhancement by magneto-mechanical resonance. J. Magn. Magn. Mater. **320**, 1553–1556 (2008)
18. D. Jie, Q. Qian, C.A. Li, *Numerical Calculation Method and the Engineering Application in Fracture Mechanics* (Science Press, Beijing, 2009)
19. A. Sarkar, P. Mukherjee, P. Barat, X-ray diffraction studies on asymmetric-call broadened peaks of heavily deformed zirconium-based alloys. Mater. Sci. Eng. **485**, 176–181 (2008)

Chapter 5
Frictional Wear Induces MMM Signals

5.1 Introduction

Apart from tensile, compression and bending loads, friction can also induce stress. Friction and wear caused by rubbing and snagging are the main failure modes of mechanical parts [1]. Approximately one-third to one-half of the world's energy is consumed in various friction types, and approximately 80% of part damage is caused by various wear forms. Therefore, to determine the residual life and ensure the remanufacturability of cores, it is desirable to detect the position of the wear scar and monitor the wear state during frictional wear before remanufacturing.

While under friction, the surface of ferromagnetic materials not only undergoes wear scarring and other forms of structural damage but is also subject to the tribo-magnetization phenomenon. The mechanism of tribo-magnetization was illustrated by Mishina et al. [2–4] based on a series of experiments on pin-on-block samples (iron and nickel) consisting of ferromagnetic materials. Magnetization was considered to start with the generation of magnetized wear elements with sizes ranging from approximately 15 nm to a few tens of nanometers, the same sizes as a magnetic single-domain particle. The transfer of wear particles formed by the wear elements under tribological actions was the key factor in tribo-magnetization. Furthermore, two different models based on the energy and material were proposed by Chang et al. [5] to clarify the physical mechanism of tribo-magnetization. An increase in surface temperature and a thermal effect were caused by frictional heat, which resulted in the random distribution of the magnetic domains and the generation of surface demagnetization. However, with the breakage of the oxide film and exposure of fresh material, the magnetic domains were driven towards the same direction by friction activation, and the geomagnetic field and surface magnetization were generated. Accordingly, several preliminary studies have been conducted on the wear process. Shi et al. [6] determined the magnetic signals on the surface of 45 steel materials before and after grinding. The magnetic signals produced abnormal magnetic peaks in the wear area, and a relationship was found between the width of the abnormal magnetic peak and the size of the wear area. Zhao et al. [7] investigated the frictional

© Science Press 2021
H. Huang et al., *Metal Magnetic Memory Technique and Its Applications in Remanufacturing*, https://doi.org/10.1007/978-981-16-1590-0_5

wear of pure iron specimens under dry friction conditions and proposed that the variation in the metal magnetic memory (MMM) signals could reflect the change in the wear forms of the specimens. The effects of the remanence of materials and the Earth's magnetic field on tribo-magnetization were also determined, and the results showed the potential for an online monitoring method based on the magnetic field for the wear state [8]. Related studies have shown the feasibility of determining the friction process and the wear state based on the variation of MMM.

Although preliminary studies have been conducted on the inspection of the wear state by MMM, few studies have been reported on the detection and analysis of surface damage and defects caused by mutual contact and relative motion. Thus, in this chapter, the variations in the tribology parameters during reciprocating sliding friction were investigated, and changes in the MMM signals along the testing directions parallel and normal to sliding were analyzed in Sect. 5.2. In addition, the single sliding friction caused by disassembly is another substantial wear form. Thus, the variations in MMM signals induced by single disassembly friction damage are also discussed in Sect. 5.3. This study can evaluate the wear degree of the damaged area on the sample surface and propose a new method for monitoring the wear state using MMM before remanufacturing.

5.2 Reciprocating Sliding Friction Damage

To investigate the variations in MMM signals under reciprocating sliding friction, linear reciprocating sliding experiments were conducted on a self-developed frictional wear tester machine of block-on-block type in the presence of the geomagnetic field. A schematic of the experimental apparatus is shown in Fig. 5.1. The upper block was installed in a block holder, and a normal load was applied to the

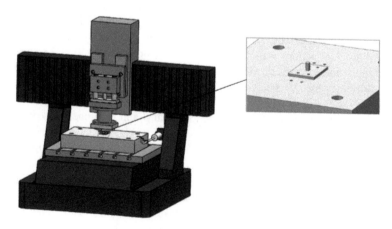

Fig. 5.1 Schematic of the experimental apparatus

upper block by a microcomputer loading system. The lower block was fixed on a moving platform made of nonferromagnetic materials to eliminate the interference of an external magnetic field during the tribo-magnetization process, and the sliding distance was controlled by a travel switch. The friction coefficient was calculated according to the frictional force measured in real time by a pressure sensor. The wear loss of the block before and after predetermined numbers of sliding cycles was measured using an AUW-220D electronic analytical balance. Moreover, a Dantsin Trimos model TR Scan 3D surface topography device was used to measure the wear depth. The magnetic signal on the surface of the lower block was measured using a TSC-2M-8 magnetic memory device with a lift-off value of 1 mm.

Typical 45 carbon steels (corresponding to ASTM 1045) were used in the reciprocating sliding experiments because of their excellent comprehensive mechanical properties and extensive industrial applications. The chemical compositions of these steels are listed in Table 5.1. The dimensions of the upper and lower blocks were constrained by the frictional tester, as shown in Fig. 5.1. The whole surfaces of both the upper and lower specimens were ground and polished to reduce the roughness of the contact surfaces to less than 0.5 μm and then cleaned with acetone before testing. Before frictional testing, scanning lines located at the center of the wear area were marked on the surface of the lower sample as shown in Fig. 5.2. The specimens were demagnetized by a TC-2 AC demagnetizer to purify the initial magnetic field on the surface.

Table 5.1 Chemical composition (wt%) of 45 carbon steels

Material	C	Mn	Si	Cr	Ni	Fe
45 steel	0.46	0.55	0.25	$\leq$0.25	$\leq$0.3	As remainder

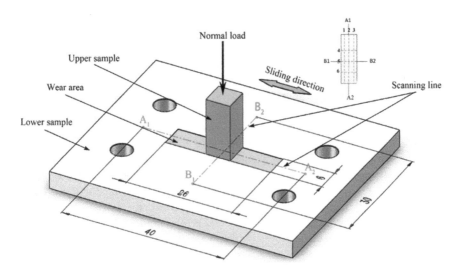

Fig. 5.2 Experimental setup: wear area and scanning lines

The sliding experiments were performed at room temperature (approximately 25 °C) under constant ambient relative humidity (approximately 50%) under dry friction conditions. A normal load of 100 N and a low sliding velocity of 0.02 m/s were applied during the experiments, and the length of the sliding stroke was 20 mm. After the predetermined number of sliding cycles was completed, the lower specimen was taken from the platform and laid in the north-south direction in a nonmagnetic environment to measure the magnetic field intensity, and the magnetic field on the wear surface was measured every 20 cycles until the end of the experiments. Each scanning line was measured three times, and the mean values of the measurements were considered the final results to reduce the errors caused by the inspection device. The mean value of the depths on three different inspection lines in the wear area, both in the directions parallel and normal to sliding, as shown in Fig. 5.2, was regarded as the wear depth.

5.2.1 Variations in the Tribology Parameters During Friction

The friction coefficient f and the wear loss m, as the two basic characteristic parameters in tribology experiments, play important roles in the analysis of the wear degree. The variations in these parameters with the number of sliding cycles during friction are shown in Fig. 5.3. In sliding cycles 0–60, as marked by A and B in Fig. 5.3, a small number of higher asperities were in contact first, and the actual contact area was small because of the low surface roughness between the contact interfaces. Thus, the friction coefficient f and wear loss m were close to zero. No visible frictional damage was observed on the surface, as shown in Fig. 5.4a. With the friction process going on, the number of asperities between the interfaces gradually increased, and the actual contact area became larger, which caused the deformation and fracturing of the asperities under normal loads. Consequently, the friction coefficient increased sharply after the 60th cycle. Accordingly, the oxide film between the contact surfaces became unstable and began to breakdown, and metal particles were generally shed from the contact surfaces [9]. The particles were pressed into the friction surface under the normal load and forced to shear and cut the surface by frictional behavior because of the furrow during sliding. This phenomenon produced obvious wear scarring and wear debris, as marked by the dashed frame in Fig. 5.4b. In addition, the wear scar width and wear loss gradually increased with the number of sliding cycles. When the number of sliding cycles reached 160, as marked by C in Fig. 5.3, the wear scar width increased to 6 mm, which matched the size of the upper specimen. With a further increase in the number of sliding cycles, the friction coefficient f fluctuated to some extent but generally showed a rising trend, while the wear loss m rapidly increased. During this stage, the oxide film was broken completely, and fresh material was exposed and destroyed by considerable wear debris. Thus, a large amount of fine debris was continuously generated and gathered at the two sides of the wear area, as marked by the dashed frame in Fig. 5.4c, which indicated that the samples were in the severe wear stage.

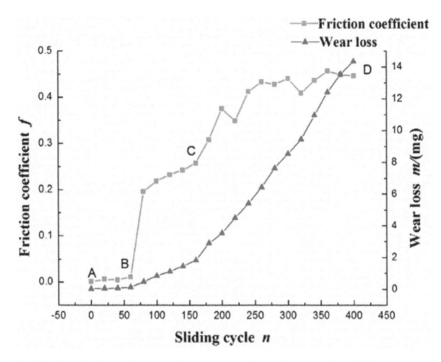

Fig. 5.3 Variations in the friction coefficient and wear loss with the number of sliding cycles

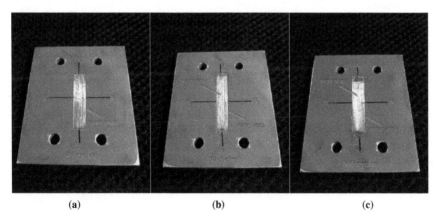

Fig. 5.4 Evolution of the wear scar on the sample surface at different stages: **a** 40 cycles; **b** 60 cycles; and **c** 160 cycles

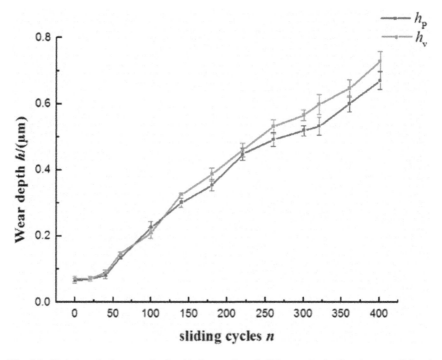

Fig. 5.5 Variations in the wear depth with the number of sliding cycles in directions parallel and normal to sliding

   In addition, variations in the wear depth with the number of sliding cycles in directions parallel and normal to sliding were investigated, as shown in Fig. 5.5. The wear depth slightly changed in the initial stage and then increased sharply with the increase in the number of sliding cycles during testing. Thus, the wear depth showed a similar tendency to the wear loss. In particular, the value of the wear depth in the vertical direction (h_v) was larger than that in the parallel direction (h_p) for a given sliding cycle. It can be concluded that these tribology parameters could be qualitatively used to test the friction process and estimate the wear degree during wear.

5.2.2 Variations in the Magnetic Memory Signals Parallel to Sliding

Figure 5.6 shows the variations in the surface magnetic signals parallel to the sliding direction with different numbers of sliding cycles. Figure 5.6a illustrates that overall, the curves of the tangential component $H_p(x)$ moved up with increasing reciprocating sliding cycles, while the curves of the normal component $H_p(y)$ moved down with

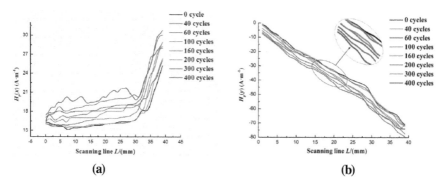

Fig. 5.6 Variations in the magnetic signals with different numbers of sliding cycles for **a** tangential component $H_p(x)$ and **b** normal component $H_p(y)$ in the direction parallel to sliding

increasing sliding cycles. Thus, the amplitudes of the magnetic signals for both $H_p(x)$ and $H_p(y)$ increased with the number of sliding cycles. This result is considered to be related to the wear degree. The integrity of the specimen surface was destroyed by reciprocating sliding, and fresh substrate materials were exposed to air, which caused plastic deformation under the friction surface. With the rubbing process, the direction of the magnetostriction magnetic domain within the wear area underwent an orientational and irreversible change under the combined action of the geomagnetic field and the normal load. This phenomenon made the magnetic domain walls deflect along the direction of the easy axis [10], resulting in the appearance of the leakage magnetic field in the stress concentration and plastic deformation areas. Moreover, the amplitude of the leakage magnetic field increased with increasing number of sliding cycles.

However, the location of the wear scar was almost impossible to determine from the curves of $H_p(x)$ and $H_p(y)$. The magnetic signals with and without debris visibly differed when the debris that had gathered at the two ends of the wear area after 400 sliding cycles was removed, as shown in Fig. 5.7 (when the specimen was at the severe wear stage). Wear debris generated from ferromagnetic materials during wear is magnetized by frictional behavior, and the magnetic field strength of the magnetized wear debris is associated with the sizes and quantities of the accumulated debris [11]. Magnetized wear debris gathered at the two ends of the wear area and particularly affected the inspection results for the magnetic signals.

The curves of $H_p(x)$ had trough features when the debris was removed, as marked by the two vertical lines in Fig. 5.7a; the locations of the vertical lines correspond to the two troughs in the curve. The gradient K of $H_p(y)$ had a visible peak when the debris was removed, as similarly marked by the two vertical lines in Fig. 5.7b. It was found that the estimated length of the wear scar was very close to the actual length. Therefore, $H_p(x)$ and K could properly locate the size of the wear area parallel to sliding after the influence of the wear debris was eliminated.

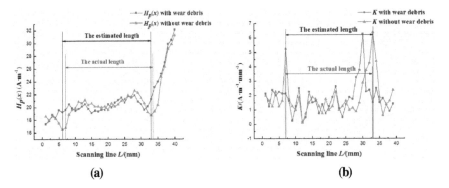

Fig. 5.7 Magnetic memory signals before and after the removal of the wear debris at 400 cycles: **a** tangential component $H_p(x)$; and **b** gradient K of normal component $H_p(y)$

5.2.3 Variations in the Magnetic Memory Signals Normal to Sliding

The distributions of $H_p(x)$ and $H_p(y)$ normal to sliding after different numbers of cycles are illustrated in Fig. 5.8. Comparing Fig. 5.4 with Fig. 5.8, changes in the magnetic signals were related to the appearance of the wear scar. Before the 60th cycle, the variations in the magnetic signals along the scanning line were modest with an increase in the number of cycles. As friction progressed, a visible wear scar appeared on the surface of the specimen, and the magnetic signals fluctuated to a certain extent. After the 160th sliding cycle, when the specimen entered the stage of severe wear, the curves of the magnetic signals showed a large variation.

$H_p(x)$ showed an obvious V-shaped distribution along the scanning line in Fig. 5.8a, and the minimum value of the V-shaped features decreased with an increase in the number of sliding cycles. $H_p(y)$ had apparent peak-trough features in the wear

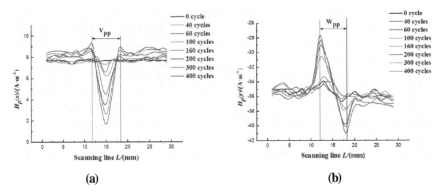

Fig. 5.8 Distributions of the magnetic signals along the scanning line after different numbers of sliding cycles: **a** tangential component $H_p(x)$ and **b** normal component $H_p(y)$

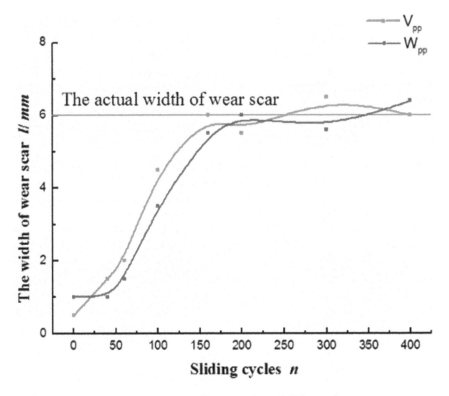

Fig. 5.9 Variations in V_{pp} and W_{pp} after different numbers of sliding cycles

area in Fig. 5.8b, and the difference value between the peak and trough increased with the number of sliding cycles. The distance between the two inflection points of the V-shaped distribution of $H_p(x)$ (as V_{pp}) and the width of the peak-to-trough feature of $H_p(y)$ (as W_{pp}) were calculated from Fig. 5.8, and the variations in V_{pp} and W_{pp} after different numbers of sliding cycles are illustrated in Fig. 5.9. V_{pp} and W_{pp} showed an increasing trend before the 160th sliding cycle. Afterwards, both parameters reached maximum values and then remained stable with an increase in the number of sliding cycles, and the two parameters were almost equal to the actual width of the wear scar. It was concluded that both $H_p(x)$ and $H_p(y)$ normal to sliding were effective indicators for estimating the width of the wear area during the severe wear stage. In the middle part of the wear area, the magnetic domains beneath the friction surface were activated and rearranged due to the normal load and sliding friction. This phenomenon resulted in the magnetization of the wear surface and an increase in the magnetic field strength. However, severe structural changes (i.e., grooves) at both ends of the wear area led to the accumulation of numerous magnetic charges at the groove interface, which substantially affected the distribution of the leakage magnetic field on the wear surface. Moreover, $H_p(x)$ of the leakage magnetic field had a trough feature, while $H_p(y)$ had a peak-trough feature. These characteristics can be used to determine the location of the wear area, as shown in Fig. 5.8.

Moreover, the maximum gradient K_{max} of $H_p(y)$ has commonly been used to characterize the degree of the stress concentration of ferromagnetic materials. K_{max} of $H_p(y)$ of the magnetic memory signal normal to the sliding direction is defined as follows:

$$K_{max} = max\left|(H_p(y)_{i+1} - H_p(y)_i)/\Delta L\right| \qquad (5.1)$$

where $H_p(y)_i$ is the magnetic signal $H_p(y)$ at coordinate i, and ΔL is the distance between adjacent inspection points along the scanning line.

The variation in K_{max} with the number of sliding cycles is illustrated in Fig. 5.10 to better understand the relationship between the magnetic signals and friction times. The amplitude of K_{max} increased slightly at the beginning, when no visible wear scarring appeared before 60 sliding cycles, and sharply increased with the number of sliding cycles once the wear scar appeared. During the frictional experiments, the wear damage and stress concentration increased with the number of sliding cycles, and the variation in K_{max} was remarkably similar to that of the friction coefficient f and the wear loss m in Fig. 5.3. Therefore, the analytical results from the evolution of

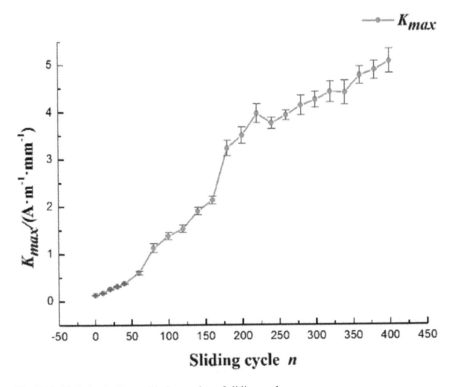

Fig. 5.10 Variation in K_{max} with the number of sliding cycles

the wear process and the variation in the magnetic signals on the sample surface show that the magnetic signal K_{max} could characterize the wear degree of the ferromagnetic materials.

5.2.4 Relationship Between the Tribology Characteristics and Magnetic Signals

Tribo-magnetization consists of two aspects, namely, the wear process accompanied by the variation in the tribology characteristics (i.e., wear loss m, friction coefficient f and wear depth h) and the magnetization process accompanied by the variation in the magnetic signals (i.e., maximum gradient K_{max} of $H_p(y)$). The experimental results showed that variations in the magnetic signals were related to the wear degree, which could be represented by the tribological characteristics. Therefore, the relationship between the tribology characteristics and the magnetic signals should be investigated further, as shown in Fig. 5.11. Figure 5.11a illustrates the wear loss m, which approximately exhibits an exponentially increasing trend with an increase in K_{max}, while the friction coefficient f shows a linearly increasing trend with K_{max}, as shown in Fig. 5.11b (initial data were ignored, similar to the following analysis). The quantitative relationships between m or f and K_{max} can be fitted by the following equations:

$$m = -0.024 + 0.214 \times \exp(K_{max}/1.162) \tag{5.2}$$

$$f = 0.117 + 0.071 \times K_{max} \tag{5.3}$$

The values of R (adjusted deviate square), which are usually used to describe the fitting quality of the model, were 0.9331 and 0.9326, respectively. Based on the same

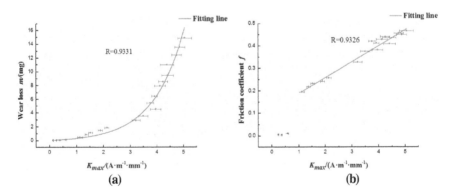

Fig. 5.11 Relationships between the tribology characteristics and magnetic signal characteristic K_{max}: **a** wear loss m; and **b** friction coefficient f

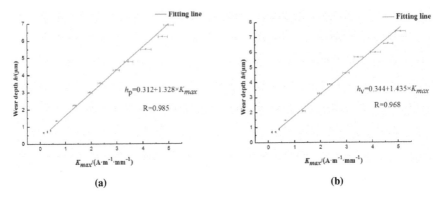

Fig. 5.12 Relationships between the wear depth and magnetic signal characteristic K_{max}: **a** h_p; and **b** h_v

method, the relationship between the wear depth and magnetic memory signals was also investigated in Fig. 5.12. A linear relationship between the wear depth h and maximum gradient K_{max} was obtained with parameters R of 0.985 for the parallel direction and 0.968 for the normal direction. These values suggest that the fitting equations were excellent, based on which changes in the tribology characteristics could be monitored by magnetic signals in a simple way.

Considering that the results to date were based on a single experiment, another experiment where the normal load was changed from 100 to 150 N was implemented to further verify the relationships between the wear degree and magnetic signal characteristic. The fitting results, which were measured by the same method, are shown in Fig. 5.13. The verification experiment demonstrates the repeatability of the experimental results, and the large R values also imply that the accuracy of the model is good. The experimental results show that the wear loss m was an exponential function of the maximum gradient K_{max}, while the friction coefficient f and wear depth h increased linearly with K_{max}. Although the parameters of the curves for 100 and 150 N differed, the quantitative relationship types between the tribology characteristics and the magnetic signals were not affected by the change in the normal load and showed good consistency between the two experiments. Thus, K_{max} may be a useful index for evaluating the damage degree of ferromagnetic materials by estimating the tribology characteristics.

5.3 Single Disassembly Friction Damage

In addition to reciprocating sliding friction, single sliding friction is another important wear form. Single sliding friction damage is generally induced during the process of disassembly for maintenance or remanufacturing. Compared with reciprocating sliding friction, the damage caused by single disassembly friction consists mainly of

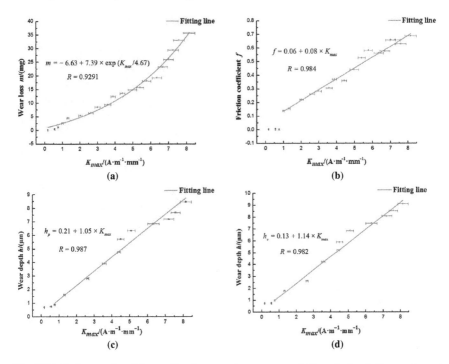

Fig. 5.13 Fitting results of the verification experiment: **a** m and K_{max}; **b** f and K_{max}; **c** h_p and K_{max}; and **d** h_v and K_{max}

scratches, furrows, and material accumulation and adhesion with no grinding debris and less mass loss. To show the difference in measured MMM signals between these two types of sliding friction, another self-developed experimental device was used to simulate the wear state of single sliding friction, as shown in Fig. 5.14a, and the test platform is shown in Fig. 5.14b. The upper specimen was made of impeller material FV520B with a size of $10 \times 10 \times 16$ mm³, and the lower specimen was made of shaft material 40CrNiMo7 with a size of $40 \times 40 \times 16$ mm³. Damage occurred when the lower specimen slid along the horizontal load F_2 direction under the action of normal load F_1. The contact pressures were set to 50 MPa, 100 MPa, 150 MPa, 200 MPa and 250 MPa, while the corresponding normal loads F_1 were set to 0.5 t, 1.0 t, 1.5 t, 2.0 t and 2.5 t, respectively. The pressure was applied to different specimens for a single sliding distance, and the sliding distance was 20 mm.

Before the experiment, the specimens were ground and polished to Ra 0.8 to reduce the roughness of the contact surfaces. As in the previous experiment, the specimens were demagnetized by a TC-2 AC demagnetizer. The MMM signals of seven marked scanning lines were measured offline. A Dantsin Trimos TR-SCAN

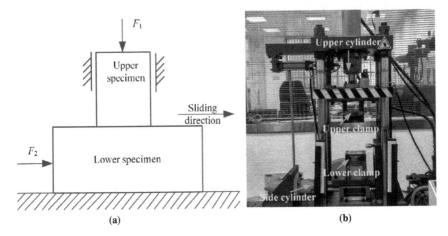

Fig. 5.14 Experimental apparatus: **a** simplified disassembly diagram and **b** test platform

Fig. 5.15 The distributions
of the scanning lines and
sliding area on the surface of
the lower specimen

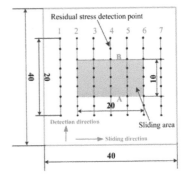

3D surface topography device, which can automatically analyze the morphological
parameters, was used to assess the topography of the sliding area. The residual
stresses on the scanning lines (from No. 1 to No. 7) were tested before and after
the disassembly experiments by a Proto-X ray residual stress analyzer. Figure 5.15
shows the distribution of the sliding area and scanning lines on the specimen surface.

5.3.1 Surface Damage and Microstructure Analysis

To investigate the formation mechanism of surface damage, the surface morphology
of the damage region (10×20 mm^2) on the disassembly interface is measured,
as shown in Fig. 5.16. After sanding and polishing, the surfaces of the upper and
lower specimens still have many micron-sized uneven asperities. The higher asperity
is immediately in contact and deforms under the normal load during disassembly.
Elastic deformation mainly occurs when the applied load is small, the asperity can be

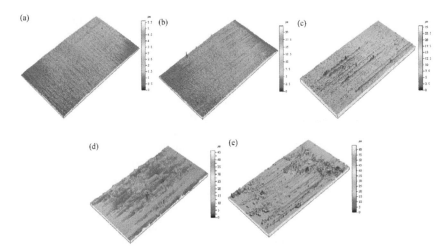

Fig. 5.16 The surface morphology of the damage region under different contact pressures: **a** 50 MPa, **b** 100 MPa, **c** 150 MPa, **d** 200 MPa, and **e** 250 MPa

restored to its original state after revoking the load, and no obvious damage appears on the contact surface, as shown in Fig. 5.16a. As the load increases to a certain extent, the pressure between the local asperity reaches or exceeds the limit of the yield strength, causing the breakage of the asperity to produce plastic deformation. Slight scratches appear as a result of the relative motion of the contact surfaces, as shown in Fig. 5.16b. The transfer particles are formed by the fractured asperity plough grooves and cracks on the surface during sliding, and the embedded particles that are pressed into the surface tear the surface to cause serious attrition as grinding grooves, as shown in Fig. 5.16c. With a further increase in the normal load, accompanied by an increase in the width and depth of the surface scratches, the sliding area is subjected to adhesion and material transfer, forming pits and bulges on the microscopic surface, as shown in Fig. 5.16d, e. It is concluded that the damage forms relate to the applied loads, and the increase in the applied loads could lead to more serious wear and deformation. The mechanism of disassembly damage, which is the result of unidirectional sliding under heavy loading, is different from that formed by conventional cyclic tribological behaviors. It is the combined effect of adhesive wear and abrasive wear with no wear debris and mass loss produced.

5.3.2 Variations in the MMM Signals

To eliminate the magnetic field interference of the ferromagnetic experimental instrument, the magnetic field H on the specimen surface after demagnetization but before loading is regarded as the initial value. The difference between the magnetic signal after loading and the corresponding initial value, marked as ΔH, is investigated. The experimental results under a contact pressure of 100 MPa are chosen to investigate the variations in the magnetic signals, as shown in Fig. 5.17.

As can be observed in Fig. 5.17, the distribution curves of the magnetic memory signals on different scanning lines are different. The magnetic signals from scanning lines 1 and 7, which are beyond the sliding area, varied almost near the zero line. The magnetic signals from the rest of the scanning lines, which are within the sliding area, exhibit substantial changes. The $\Delta H_p(x)$ components of magnetic signals on scanning lines 2–6 show visible peak and trough features at the locations of 5 mm and 15 mm, respectively, as well as zero-crossing points at the location of 10 mm, as shown in Fig. 5.17a. In contrast, variations in the $\Delta H_p(y)$ components have evident peak features with a maximum value at the location of 10 mm on scanning lines 2 to 6, as shown in Fig. 5.17b. The magnetic signals have distinct distortions at locations from 5 mm to 15 mm, which corresponds exactly to the sliding area shown in Fig. 5.15; in addition, the change in the magnetic signals beyond the sliding area is not obvious. This phenomenon indicates that the magnetic memory signals can be used to locate the sliding area.

Generally speaking, the leakage magnetic signal excited by the combined actions of the geomagnetic field and mechanical load in the stress concentration zones can be characterized by the tangential component $H_p(x)$ having a maximum and normal component $H_p(y)$ having a value of zero and intensively changing polarity. It is worth noting, however, that the features of the magnetic signals measured in the disassembly experiment are completely contrary to those measured in previously published studies. To explore the generation mechanism of the surface magnetic signal in disassembly, the residual stresses of each detection point on the scanning lines from 1 to 7 are measured by a Proto-X ray residual stress analyzer.

The difference value ΔP of the residual stresses before and after the disassembly experiment is calculated, as shown in Fig. 5.18a. The residual stresses on scanning lines 1 and 7 are essentially zero, while there is an apparent increase at locations from 5 to 15 mm on scanning lines 2–6. This means that the stresses experience a dramatic change within the sliding area but basically remain unchanged beyond the sliding area. The amplitude of ΔP on scanning lines 2 to 6 gradually increases first

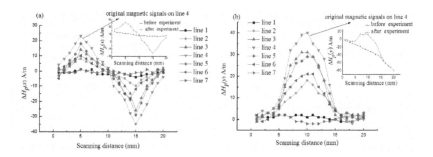

Fig. 5.17 Variations in the magnetic memory signals under a contact pressure of 100 MPa: **a** tangential component $\Delta H_p(x)$ and **b** normal component $\Delta H_p(y)$

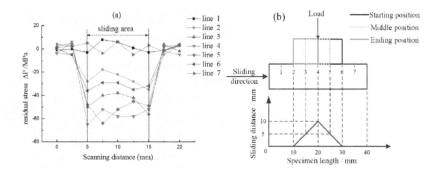

Fig. 5.18 Variations in the residual stress on each detection line under a contact pressure of 100 MPa: **a** the difference value ΔP, and **b** the sliding distance at different scanning lines

and then decreases as a whole along the sliding direction and reaches a maximum at middle line 4. This may be interpreted as shown in Fig. 5.18b. The relative sliding distance between the upper and lower specimens reaches the maximum at line 4 and the minimum at lines 1, 2, 6 and 7. This means that the adhesive and abrasive wear degree on the middle part of the specimen is more serious than that on the edge part. As a result, the change in the residual stresses on line 4 is larger than that on lines 1, 2, 6, and 7.

Under the effect of contact pressure, plastic deformation and stress concentrations appear at both sides of the sliding area, marked as A and B in Fig. 5.15, during the disassembly process. The discontinuity of the material and the distortion of the structure are mainly concentrated at sides A and B because of plastic deformation, as shown in the schematic diagram of the side profile of the sliding area in Fig. 5.19.

It can be assumed that there is a set of linear magnetic charges at sides A and B. Be different from the traditional magnetic charge distribution as shown in Fig. 3.2c, in the single disassembly friction damage condition, the bulge of the plastic deformation can induce positive magnetic charges, while the trough of the sliding area induces negative magnetic charges. Based on magnetic charge theory, each component of the

Fig. 5.19 Schematic diagram of the side profile of the sliding area

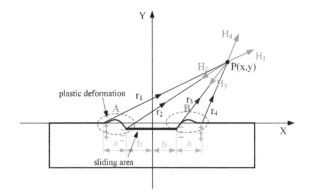

magnetic field at the space point $P(x, y)$ caused by magnetic charges can be expressed as:

$$\mathbf{H}_n = \frac{\rho}{2\pi \mu_0 r_n^2}\mathbf{r}_n \qquad n = 1, 2, 3, 4 \qquad (5.4)$$

where ρ is the magnetic charge density, μ_0 is the permeability of a vacuum, and r_n represents the distance between point P and the magnetic charge. Then, the tangential component of the magnetic field $H_p(x)$ and the normal component $H_p(y)$ of the magnetic field can be given as follows:

$$H_P(x) = \frac{\rho}{2\pi \mu_0}\left[\frac{x+a+b}{(x+a+b)^2+y^2} - \frac{x+b}{(x+b)^2+y^2}\right.$$
$$\left. - \frac{x-b}{(x-b)^2+y^2} + \frac{x-a-b}{(x-a-b)^2+y^2}\right] \qquad (5.5)$$

$$H_P(y) = \frac{\rho}{2\pi \mu_0}\left[\frac{y}{(x+a+b)^2+y^2} - \frac{y}{(x+b)^2+y^2}\right.$$
$$\left. - \frac{y}{(x-b)^2+y^2} + \frac{y}{(x-a-b)^2+y^2}\right] \qquad (5.6)$$

where a is the width of the linear magnetic dipole and b is the half width of the sliding area. Without loss of generality, the parameters in the magnetic charge model can be set as $\rho = 1.0 \times 10^{-5}$ N/A, $\mu_0 = 4\pi \times 10^{-7}$ N/A^2, $a = 8$ mm, and $b = 5$ mm. The calculated curves are shown in Fig. 5.20. The variations in the theoretical results are consistent with the experimental measurements, as shown in Fig. 5.17.

Although the amplitudes of the tangential and normal component signals are slightly different on each scanning line, the change laws are similar, as shown in Fig. 5.17. ΔH of the middle scanning line in each test is chosen to further analyze the

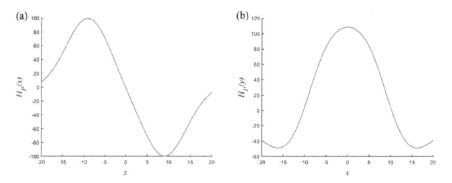

Fig. 5.20 Curves with numerical calculation: **a** tangential component $H_P(x)$ and **b** normal component $H_P(y)$

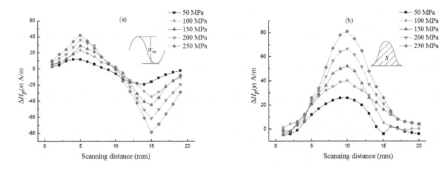

Fig. 5.21 Distributions of the magnetic signals on the middle detection line under different loads: **a** tangential component $\Delta H_p(x)$ and **b** normal component $\Delta H_p(y)$

variations in the magnetic memory signals under different loads during disassembly, as shown in Fig. 5.21.

The magnetic signals ΔH on each middle scanning line show the same change law, namely, the curves of tangential component $\Delta H_p(x)$ exhibit obvious peak and trough features at the sides of the sliding area, while the curves of normal component $\Delta H_p(y)$ show peak features within the sliding area. The correlated features, as shown in Figs. 5.17 and 5.21, indicate that the magnetic memory signal can be utilized to quickly detect disassembly defects. In addition, both the peak-to-trough difference W_{pp} of the tangential component and the peak area S of the normal component increase with increasing load. The larger mechanical loads intensify the plastic deformation and stress concentration, which aggravate the deviation and rotation of the magnetic domains, and the surface magnetization increases as a result. Further analysis indicates that the distorted magnetic signals $\Delta H_p(x)$ and $\Delta H_p(y)$ are capable of locating disassembly defects and qualitatively evaluating the damage degree.

5.3.3 Damage Evaluation of Disassembly

From the above analysis, it can be concluded that scratches, grooves and adhesive wear surface damage are easily and inevitably generated when mechanical loading is applied. The degree of surface damage, which depends on the applied load, has a substantial impact on the process of remanufacturing inspection and repair. Therefore, the average damage depth Sa is regarded as an index for evaluating the surface damage degree. It is defined as the arithmetic mean of the absolute of the ordinate values within the definition area, and the formula is shown as follows:

$$Sa = \frac{1}{a} \int \int |\eta(x, y)| dx dy \approx \frac{1}{MN} \sum_{i=1}^{M} \sum_{j=1}^{N} |\eta(i, j)| \tag{5.7}$$

where M and N represent the number of sampling points in the X and Y directions, respectively, and $\eta(i, j)$ represents the height of the roughness irregularities from the mean value.

The average damage depth Sa is extracted from the surface morphology of the 10×20 mm^2 damage region, which is detected by Dantsin Trimos model TR-SCAN 3D surface topography device. The variation between the average damage depth and the contact pressure is established by taking the load as the independent variable and the damage depth as the dependent variable, as shown in Fig. 5.22. The average damage depth shows an increasing trend with increasing contact pressure. Specimens with different damage depths need to be remanufactured by means of different surface repair processes. When a small load is applied, the surface damage after disassembling is small, which can be continued to use without repair, as shown in Fig. 5.16a. With the increase in loads, mild scratches and adhesive wear appear in the local area of the surface as indicated in Fig. 5.16b, c, which can be upgraded by means of electric brush plating. Surface damage is further exacerbated when obvious furrows and large areas of adhesive form on the surface as shown in Fig. 5.16d, e, and heavy repair work should be implemented by means of laser cladding or plasma spray welding. A further increase in the contact pressure will lead to serious furrows and exfoliation and accumulation of metal particles, resulting in a sharp increase in

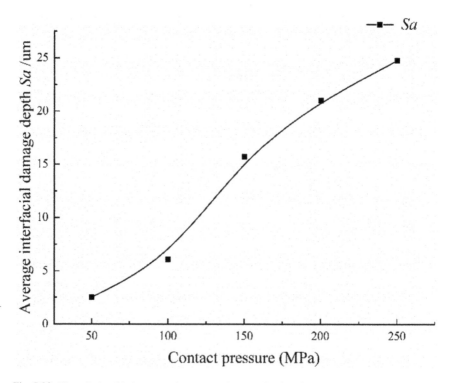

Fig. 5.22 The relationship between the average damage depth and contact pressure

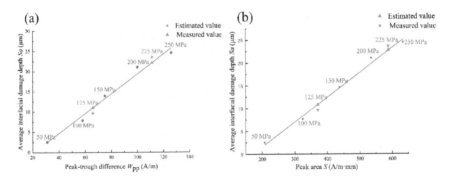

Fig. 5.23 The relations between the average interface damage depth Sa and magnetic signal characteristics: **a** peak-trough difference W_{pp}, and **b** peak area S

the remanufacturing repair cost or a repair scenario that is even beyond feasibility. From the view of remanufacturing, the parameter Sa directly affects the selection of repair processes; thus, the parameter Sa is a reasonable index to quantitatively evaluate the degree of interface damage.

The detection of the average damage depth Sa requires an exclusive 3D surface topography device, which has the shortcomings of complex operation, time consumption and unfavourability in engineering. Magnetic memory signal characteristics can qualitatively assess the disassembly damage degree based on the above analysis. Therefore, the quantitative relationship between the damage degree and magnetic signal characteristics is further discussed, as shown in Fig. 5.23. Linear equations between the average damage depth Sa and W_{pp} and S are obtained based on the least squares method. The results are as follows:

$$Sa = 0.244 \cdot Wpp - 5.1 \tag{5.8}$$

$$Sa = 0.054 \cdot S - 9.3 \tag{5.9}$$

The correlation coefficients R of the fitting results are 0.975 and 0.986, respectively. The large correlation coefficients indicate that the error between the fitting results and the experimental results is acceptable. This proves that the average damage depth Sa can be predicted by measuring the magnetic signals in a simple and time-saving way.

Two additional experiments are implemented to further verify the accuracy of the relations. The contact pressures in the verification experiments are set at 125 and 225 MPa. The Sa value obtained from measurements is compared with the estimated value Sa calculated by Eqs. (5.8) and (5.9), as shown in Fig. 5.23. The differences between the measured value and the estimated value of Sa are small and are within a permissible range, which demonstrates that the relationships between Sa and W_{pp} and Sa and S are reasonable and repeatable. Therefore, the nondestructive evaluation of the disassembly damage can be achieved by predicting the damage depth on the basis of metal magnetic memory technology, and it also provides a reference for the remanufacturing performance and selection of repair processes.

5.3.4 Verification for Feasibility and Repeatability

To verify the feasibility of the proposed magnetic memory method in disassembly damage evaluation, the eddy current testing method was further conducted. During the test, the eddy current signals on each scanning line of the lower specimen were measured by OLYMPUS NORTEC 600 eddy current flaw detector combined with a 7.9 mm diameter U8626004 spot probe. The signals were compared with those measured from the reference standard specimen, as shown in Fig. 5.24.

The coil impedance plane resistance of the lower specimen and reference standard specimen is illustrated in Fig. 5.25. The eddy current signals on the lower specimen remain stable, and no noticeable variation in the signals can be noted. This implies that eddy current testing may not be appropriate for disassembly damage detection. One explanation is that the sensitivity of the probe is not sharp enough to distinguish the defects. The slightest defect depth that can be identified by the probe is 0.08 mm, while the damage depth of the lower specimen was 5–60 μm. Under such circumstances, the magnetic memory method can be adopted.

To evaluate the repeatability of magnetic memory methods for disassembly damage detection, contrast tests were conducted with five additional samples. During the tests, the contact pressure was set at 50, 100, 150, 200, and 250 MPa. The pretreatment process and the measurement method were consistent with the experiments aforementioned. Variations in the magnetic memory signals under a contact pressure of 150 MPa can be seen in Fig. 5.26. The tangential components of the magnetic signal $\Delta H_p(x)$ have peak-trough features with zero-crossing points, and this feature at the middle position can be observed more obviously, as shown in Fig. 5.26a.

Fig. 5.24 The distributions of the scanning lines on the reference standard specimen

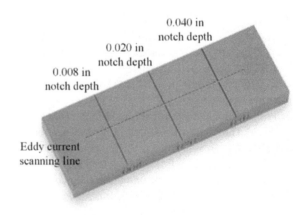

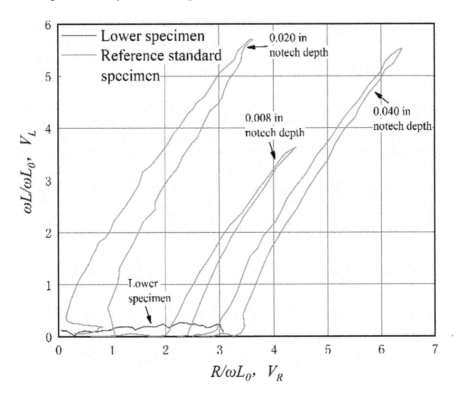

Fig. 5.25 Coil impedance on both the lower specimen and reference standard specimen

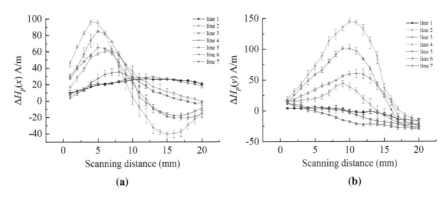

Fig. 5.26 Variations in magnetic memory signals under a contact pressure of 150 MPa: **a** tangential component $\Delta Hp(x)$ and **b** normal component $\Delta Hp(y)$

The normal components of the magnetic signal $\Delta H_p(y)$ show peak features, and the maximum value appears at line 4, as shown in Fig. 5.26b. These variation trends of the magnetic memory signals in the contrast tests are similar to those under a contact pressure of 100 MPa, as shown in Fig. 5.17.

ΔH on the middle scanning line (line 4) under different loads was plotted in Fig. 5.27. The curves of tangential component $\Delta H_p(x)$ exhibit obvious peak-trough features, while the curves of normal component $\Delta H_p(y)$ show peak features. Furthermore, both the characteristic values of the tangential component and normal component, namely, the peak-trough difference W_{pp} and peak area S, increase with increasing load. Although the amplitudes were different, the change trends of the curves were similar to those shown in Fig. 5.21.

The average damage depth Sa was extracted from the damage region, as shown in Fig. 5.28. With increasing contact pressure, Sa showed an increasing trend. There was also an approximate linear relationship among the average damage degree Sa,

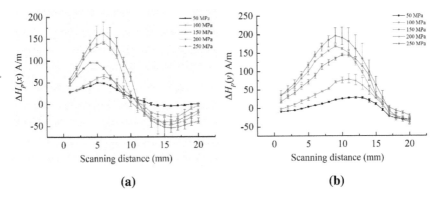

Fig. 5.27 Magnetic signals on line 4 under different loads of repeated experiments: **a** tangential component $\Delta H_p(x)$ and **b** normal component $\Delta H_p(y)$

Fig. 5.28 The relationship between the average damage depth and contact pressure of repeated experiments

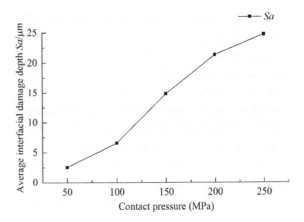

magnetic signal characteristics peak-trough difference W_{pp}, and peak area S, as shown in Fig. 5.29. The fitting results were $Sa = 0.141 \times W_{pp} - 5.43$ and $Sa = 0.018 \times S - 6.72$, and the correlation coefficients R were 0.977 and 0.939, respectively, based on calculations with the least squares method. These results were similar to those shown in Fig. 5.23.

The magnetic signals of each sample were measured immediately after the disassembly experiment and then measured every 24 h during the next five days. The magnetic signals measured 5 days after the experiment under a contact pressure of 150 MPa are presented in Fig. 5.30. It was found that the differences in the magnetic signals measured five days after the experiment and those obtained immediately (Fig. 5.26) were not substantial. Figure 5.31 shows the variations in the characteristics W_{pp} and S over time under a contact pressure of 150 MPa. The characteristics fluctuated with time in a certain range, but no obvious change was observed. This may

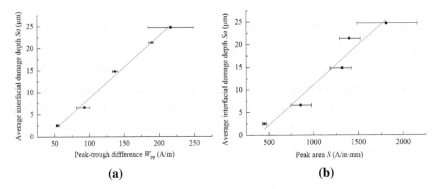

Fig. 5.29 The relations between the average interface damage depth Sa and magnetic signal characteristics of repeated experiments: **a** peak-trough difference Wpp, and **b** peak area S

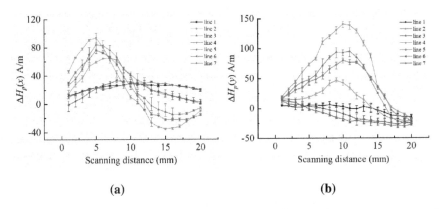

Fig. 5.30 Magnetic signals measured 5 days after the experiment under a contact pressure of 150 MPa: **a** tangential component $\Delta H_p(x)$ and **b** normal component $\Delta H_p(y)$

Fig. 5.31 Variations in the characteristics W_{pp} and S over time under a contact pressure of 150 MPa

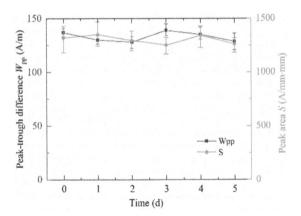

be because the release of residual stress occurs throughout months or even years in the natural state, while this experimental interval was only a few days, which requires further study.

It should be noted that only the magnetic signals on line 4 were analyzed in all the experiments where the signal strength was the largest. The linear relationship on other scanning lines can also be established; however, the magnetic signals are affected by many factors, such as the external stray magnetic field and a uniform stress state, which require further study.

5.4 Conclusions

An approach of measuring the MMM signals has been proposed to rapidly locate the wear area and characterize the wear degree for ferromagnetic materials based on tribo-magnetization. It has been observed that the tribology characteristics (i.e., friction coefficient f, wear loss m and wear depth h) are dependent on the evolution of the surface wear scar. Variations in the magnetic field were found to be related to the appearance of surface damage during friction. The distance between the trough features of the $H_p(x)$ curve and that between the peak features of the gradient K curve after the wear debris was removed could be used to locate the length of the wear area parallel to sliding at the severe wear stage. Moreover, the visible V-shaped features of $H_p(x)$ and the peak-trough features of $H_p(y)$ normal to sliding could be used to characterize the width of the wear area. Additionally, the maximum gradient K_{max} of $H_p(y)$ normal to the sliding direction increased with the number of sliding cycles. This parameter could be an effective index to estimate the wear degree of ferromagnetic materials. Thus, by analyzing the relationship between the tribology characteristics and the magnetic signals, the wear loss m was found to increase exponentially with the maximum gradient K_{max}, while the friction coefficient f and wear depth h increased linearly with K_{max}. According to these relationships, the wear loss, friction coefficient and wear depth can be estimated based on K_{max}, which can be easily measured offline. Therefore, MMM can be regarded as a feasible and simple method for determining the wear state and monitoring the wear failure of ferromagnetic materials.

In addition, the feasibility of evaluating the single sliding friction damage induced by disassembly based on MMM has also been investigated. The stress concentration and plastic deformation within the damage region lead to the distortion of the surface magnetic flux signals. The changes in the $\Delta H_p(x)$ components of the magnetic signals show peak-trough features with zero crossing, and the $\Delta H_p(y)$ components have obvious peak features. The distorted signals of both the tangential and normal components are capable of locating the damage region and qualitatively evaluating the disassembly damage. The surface topography parameter Sa is regarded as the average damage depth to quantify the disassembly defects. There is an approximately linear relationship among the average damage depth Sa, peak-trough difference W_{pp} of the tangential component, and peak area S of the normal component. This indicates that disassembly damage can be quickly determined by measuring the MMM signals in an easy and time-saving way.

This chapter presents a nondestructive and effective MMM method to evaluate friction wear and provides a reference for the remanufacturability evaluation and selection of repair technologies for remanufacturing. The quantitative relationships were obtained on the basis of the obvious wear scar circumstances under laboratory conditions, and the fitting equations indicated that the wear degree could be effectively evaluated by measuring the magnetic flux leakage signals. Nevertheless, the above results only reflect the analytical conclusions under laboratory conditions, and the fitting equations were not applicable in the mild wear stage. Therefore, an accurate evaluation model for the wear state of ferromagnetic materials using magnetic signals requires further study.

References

1. I.M. Hutchings, *Tribology: Friction and Wear of Engineering Materials* (Edward Arnold, London, 1992)
2. H. Mishina, H. Iwase, A. Hase, Generation of wear elements and origin of tribo-magnetisation phenomenon. Wear **269**(5), 491–497 (2010)
3. H. Mishina, A. Hase, T. Nakase et al., Mechanism of surface magnetisation by friction of ferromagnetic materials. J. Appl. Phys. **105**(9), 093911-1-5 (2009)
4. H. Mishina, Magnetisation of ferromagnetic material surfaces by tribological process. J. Appl. Phys. **99**(11), 6721–6727 (2002)
5. Y.P. Chang, J.P. Yur, L.M. Chu et al., Effects of friction on tribomagnetisation mechanisms for self-mated iron pairs under dry friction condition. Proc. Instit. Mech. Eng. Part J: J. Eng. Tribol. **223**(6), 859–869 (2009)
6. C.L. Shi, S.Y. Dong, B.S. Xu et al., Research on metal magnetic memory test in process of frictional wear. J. Mater. Eng. **30**(4), 35–44 (2009)
7. K.P. Zhao, J.C. Fan, F.M. Gao et al., Research on tribo-magnetisation phenomenon of ferromagnetic materials under dry reciprocating sliding. Tribol. Int. **92**, 146–153 (2015)

8. F.M. Gao, J.C. Fan, Research on the effect of remanence and the earth's magnetic field on tribo-magnetisation phenomenon of ferromagnetic materials. Tribol. Int. **109**, 165–173 (2017)
9. Q. Zhang, Z.Y. Jiang, D.B. Wei et al., Interface adhesion during sliding wear in cast iron after hot deformation. Wear **301**(1), 598–607 (2013)
10. A. Hase, H. Mishina, Magnetisation of friction surfaces and wear particles by tribological processes. Wear **268**(1), 185–189 (2010)
11. F.M. Gao, J.C. Fan, K.P. Zhao et al., In-situ observation of the magnetic domain in the process of ferroalloy friction. Tribol. Int. **97**, 371–378 (2016)

Chapter 6
Stress Concentration Impacts on MMM Signals

6.1 Introduction

From previous studies, we find that whether the magnetomechanical effect is caused by typical loads, such as tension, compression and bending, or whether the friction magnetization is caused by single and reciprocating sliding friction, in essence, the spontaneous magnetic flux leakage phenomenon is affected by stress concentration. Stress concentration in these ferromagnetic components, which comes from the microscopic (gas holes, inclusions, cracks, etc. appearing during manufacturing) or macroscopic (a variety of design features, such as holes, flanges and shoulders) discontinuities of materials will lead to damage as well [1]. Therefore, investigations of stress concentration and its impact on metal magnetic memory (MMM) signals in the fatigue process have substantial meaning for the evaluation of the damage degree in remanufacturing cores.

A large number of researchers explored the physical mechanism of this spontaneous magnetic phenomenon and the relationship between MMM signal and stress concentration. Shi et al. [2] studied the relationship between the stress concentration coefficient and the MMM signal in a dynamic tension environment, and the results showed that the stress concentration coefficient was positively correlated with the absolute value of the gradient of MMM signal. Bao et al. [3] established a relationship between MMM signal under tension and residual stress and stress concentration and determined the linear relationship between the maximum gradient of the MMM signal component and the stress concentration coefficient, which can be used for the quantitative evaluation of the stress concentration. Zhong et al. [4] studied the relationship between the stress concentration and magnetic signals in different magnetic environments and found that an environmental magnetic field would change the effects of magnetic signals caused by stress concentration. The results showed that multiple tests under different environmental magnetic fields were helpful to evaluate the degree of stress concentration. In addition, they also simulated the magnetic field changes caused by stress concentration [5]. Compared with the experimental results, it is found that the normal component of the MMM signal is related to the stress

© Science Press 2021
H. Huang et al., *Metal Magnetic Memory Technique and Its Applications in Remanufacturing*, https://doi.org/10.1007/978-981-16-1590-0_6

distribution. Shi et al. [6] extended the two-dimensional magnetic charge model to the three-dimensional situation and simulated the stress concentration area with the model. These studies show that the MMM signal characteristic value will increase as the stress concentration factor increases in most cases. However, how stress concentration and microdefects impact the MMM signal of ferromagnetic structural steels still needs to be investigated in a quantitative way. Besides, it is also significant to determine the stress concentration zone and morphology parameters of remanufacturing cores accurately in order to confirm whether the cores meet the requirements of remanufacturability. The various inverse models and methods, such as the neural network [7], machine learning [8] and the optimization algorithms [9], have been established by many researchers. These research results show that the inverse problem and the forward problem of quantitation evaluation of stress concentration are closely related with each other and it is worth investigating further.

In this chapter, a possible method for testing the stress concentration degree is proposed based on both experiment and theoretical analysis. First, a dipole model was established, and the MMM signals of specimens with different stress concentration factors were detected and analyzed in detail in Sect. 6.2. In addition, the variation in the MMM signals induced by the local plastic deformation in ferromagnetic materials was discussed by proposing a new dual-dipole model in Sect. 6.3. Finally, the inversion of stress concentration was analyzed in Sect. 6.4. These studies provide a novel method for quantitatively inspecting stress concentrations in MMM testing.

6.2 Stress Concentration Evaluation Based on the Magnetic Dipole Model

6.2.1 Establishment of the Magnetic Dipole Model

Many MMM experiments and applications show that no abnormal magnetic signals can be detected until some microscopic defects have been developed on the surface of the structure or materials have been made disconnected. Therefore, the appearance of microdefects is a necessary condition for the application of the MMM method. A rectangular defect zone of the specimen with a width of $2b$ mm and a depth of h mm is prefabricated. In view of ferromagnetic interaction energy and thermodynamic equilibrium, when a defect comes into being, stress releases quickly, and demagnetization increases rapidly at the same time to maintain the balance of energy states; therefore, opposite magnetic charges accumulate on either side of the notch, as shown in Fig. 6.1. Two magnetic charge planes with the same magnetic charge density ρ_m and opposite magnetic poles are formed. The magnetic dipole model can be considered to be the simplified 2D stress concentration zone magnetic charge model (Model B in Fig. 2.8), as introduced in Chap. 2. Therefore, the surface magnetic leakage field at the space point $P(x, y)$ can be expressed as given in Eq. (6.1) based on the magnetic dipole model:

Fig. 6.1 Surface magnetic
signal at a space point P(x, y)
based on an equivalent plane
dipole model

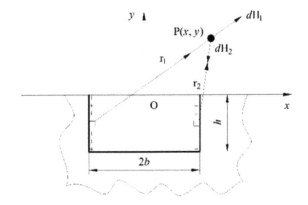

$$\begin{cases} d\mathbf{H}_1 = \frac{\rho_m \cdot d\eta}{2\pi \mu_0 r_1^2} \mathbf{r}_1 \\ d\mathbf{H}_2 = \frac{\rho_m \cdot d\eta}{2\pi \mu_0 r_2^2} \mathbf{r}_2 \end{cases} \qquad (6.1)$$

where $r_1 = \sqrt{(x+b)^2 + (y+\eta)^2}$, $r_2 = \sqrt{(x-b)^2 + (y+\eta)^2}$, $d\eta$ is the width of
the magnetic charges, and μ_0 is the permeability in a vacuum. Then, the normal
component of the magnetic leakage field $H_p(y)$ at P(x, y) can be expressed as given
in Eq. (6.2).

$$\begin{aligned} H_p(y) &= \int_{-h}^{0} d H_{1y} + \int_{-h}^{0} d H_{2y} \\ &= \int_{-h}^{0} \frac{\rho_m(y+\eta)d\eta}{2\pi \mu_0[(x+b)^2 + (y+\eta)^2]} + \int_{-h}^{0} \frac{(-\rho_m)(y+\eta)d\eta}{2\pi \mu_0[(x-b)^2 + (y+\eta)^2]} \\ &= \frac{\rho_m}{4\pi \mu_0}\left[\ln \frac{(x+b)^2 + (y+h)^2}{(x+b)^2 + y^2} - \ln \frac{(x-b)^2 + (y+h)^2}{(x-b)^2 + y^2} \right] \end{aligned} \qquad (6.2)$$

The numerical curves of $H_p(y)$ are shown in Fig. 6.2, describing how the magnetic
signals are theoretically affected by the size of the defect and the lift-off value in
the measurement. All the results are normalized by dividing the amplitudes with a
coefficient of $\rho_m/4\pi \mu_0$. The amplitude of the $H_p(y)$ signals increases with increasing
crack width b or crack length h, as shown in Fig. 6.2a, b. The lift-off value y is also an
important factor in the tests; the amplitude decreases quickly with increasing lift-off
value, as shown in Fig. 6.2c.

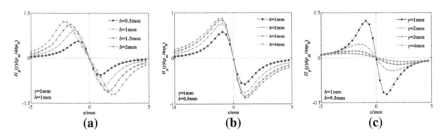

Fig. 6.2 Impact of the **a** crack width b, **b** crack depth h and **c** lift-off value y on the normal component of the leakage magnetic field

6.2.2   Characterization of the Stress Concentration Degree

The specimen was made of Q345 low-carbon steel. Its yielding strength was 358 MPa; the ultimate strength was 484 MPa; the elastic modulus was 206 GPa; the shear modulus was 79.38 GPa; and the Poisson's ratio was 0.25–0.30. The shapes of the sheet specimens and three scanning lines are shown in Fig. 6.3, and the dimensions of the precut notch for the specimens with different stress concentration factors are listed in Table 6.1. The specimens were polished and demagnetized before loading to eliminate their initial magnetic field. Dynamic tension loads were applied to the specimens on an MTS810 servo hydraulic testing machine, and the dynamic load error was within ±1.0%. Tension-tension fatigue tests with a constant amplitude (sinusoidal waveform) were performed, with the maximum load at 35 kN (stress at 328 MPa), minimum load at 3.5 kN (stress at 32.8 MPa) and load frequency at

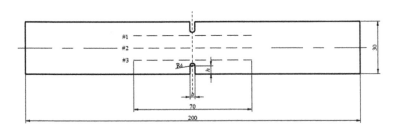

Fig. 6.3 Shape of the sheet specimens (in mm): h is the depth of the notch, b is the width of the notch and R_d is the radius

Table 6.1 Dimension of the precut notch of specimens with different stress concentration factors K_t

Specimen	(a)	(b)	(c)
K_t	3	4	5
R_d (mm)	1.5	0.8	0.5
b (mm)	3	1.6	1
h (mm)	4.6	5.3	5.6

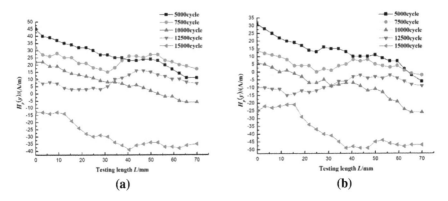

Fig. 6.4 The $H_p(y)$ distribution for specimen (a) on its **a** scanning line #1 and **b** scanning line #2

10 Hz. The experiments were carried out at room temperature. The $H_p(y)$ values were measured by an EMS 2000+ metal magnetic memory device in an Earth magnetic field environment. The probe with a 1 A/m sensitivity based on the Hall sensor was gripped on a nonferromagnetic 3D electric scanning platform and was placed vertical to the surface of the specimen with a lift-off value of 1 mm. In the fatigue testing, the specimens were unloaded when the loading cycle reached 5000, 7500, 10,000, 12,500 and 15,000. When loaded to the preset cycle number, the specimens were carefully taken from the holders and laid on the platform away from the testing machine in the south to north direction, and the $H_p(y)$ signals at the three scanning lines were measured for each specimen. Furthermore, the length of the fatigue crack was measured by a JXD-250B reading microscope during the fatigue tests.

At 15,000 loading cycles, cracks were initiated at the position of the notch for specimens (b) and (c), while no cracks were detected in specimen (a). The variations in the magnetic signals measured on lines #1, #2 and #3 have identical features at each loading cycle except for the signal amplitudes; therefore, only the results from lines #1 and #2 are presented. Figure 6.4 shows the relationship between the magnetic memory signals on the scanning lines and the loading cycles for specimen (a). It can be seen that the $H_p(y)$ value on both scanning lines decreased with the increase in loading cycles; all the $H_p(y)$ curves show a trend of declining along the direction of the scanning line; meanwhile, some fluctuations were observed at each $H_p(y)$ curve. According to the model developed by Jiles et al. [10], the magnetization is reduced with applied cyclic stress to overcome the internal friction forces so that it can approach its anhysteretic state, which is an irreversible process. This phenomenon was also investigated in both fatigue experiments and static tensile tests. Figure 6.5 and Fig. 6.6 show the variation in magnetic memory signals for specimens (b) and (c), respectively. There is no substantial change in the magnetic memory signal at the scanning lines until the loading cycle reaches 15,000 when cracks are initiated in the specimens, which is different from specimen (a) in the middle stage of fatigue. It can be observed that $H_p(y)$ varied intensively and changed its polarity in the notch area (at 35 mm) at a loading cycle of 15,000. The $H_p(y)$ curve on scanning line #1

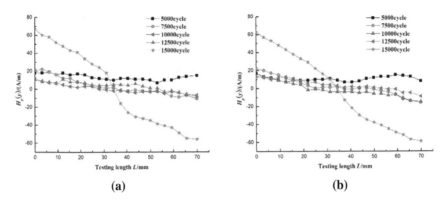

Fig. 6.5 The $H_p(y)$ distribution for specimen (b) on its **a** scanning line #1 and **b** scanning line #2

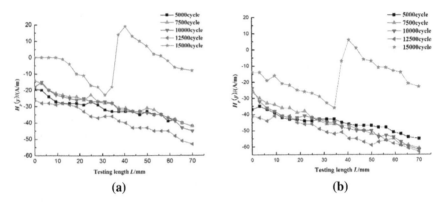

Fig. 6.6 The $H_p(y)$ distribution for specimen (c) on its **a** scanning line #1 and **b** scanning line #2

has the same variation trend as scanning line #2, except that the peak-to-trough value of $H_p(y)$ at line #1 is larger than that at line #2 for specimen (c).

To further describe the variation in magnetic signals, the magnetic gradient K was analyzed. Previous research shows that the magnetic gradient is an important parameter for describing the degree of stress concentration. The values and distributions of the gradients show both a good qualitative and quantitative correlation with the values and distributions of residual stress, and the quantitative relationships between the gradients and equivalent residual (von Mises) stress were developed and verified [11, 12]. Here, the maximum magnetic gradient K_{max} of $H_p(y)$ on the scanning line was investigated, which always appeared at the area of micro defects. The gradient K_{max} appeared at the crack area in the tests when the loading cycle reached 15,000, as shown in Figs. 6.7 and 6.8a. The theoretical value of K was given in Eq. (6.3) based on the given $H_p(y)$ in Eq. (6.2).

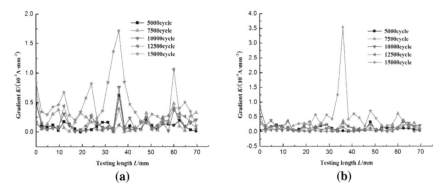

Fig. 6.7 The gradient K of magnetic memory signal on scanning line #1 of **a** specimen (b) and **b** specimen (c)

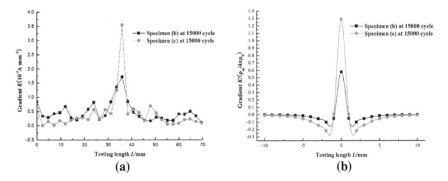

Fig. 6.8 The gradient K **a** on scanning line #1 of the specimens and **b** determined by Eq. (6.3)

$$K = \frac{\partial H_p(y)}{\partial x} = \frac{\rho_m}{4\pi\mu_0} \left[\frac{2(x+b)}{(x+b)^2 + (y+h)^2} - \frac{2(x+b)}{(x+b)^2 + y^2} \right.$$
$$\left. - \frac{2(x-b)}{(x-b)^2 + (y+h)^2} + \frac{2(x-b)}{(x-b)^2 + y^2} \right] \quad (6.3)$$

The theoretical values of K of specimens (b) and (c) are shown in Fig. 6.8b, given the crack length and width at 15,000 cycles detected from the JXD-250B reading microscope with a measurement accuracy of 0.015 mm. The crack length and width are 2.001 mm and 0.168 mm for specimen (b) and 4.121 mm and 0.387 mm for specimen (c), respectively. The measured value of $K_{max}(c)/K_{max}(b)$ (which is the ratio of K_{max} at specimen (c) to specimen (b)) was nearly equal to the theoretical ratio determined by Eq. (6.3), which was 2.5 in the tests. This implies that the maximum magnetic gradient could be an important indicator for determining the degree of stress concentration and the size of cracks or defects as well.

To provide a better understanding of the impact of stress concentration on the $H_p(y)$ signal and its gradient K, Fig. 6.9a presents the distribution of normalized gradients with crack lengths from 0.01 to 3 mm, a width b at 0.5 mm and a lift-off value y at 1 mm. The gradient K was normalized by dividing the amplitudes with a coefficient of $\rho_m/4\pi\mu_0$. Clearly, the theoretical results are coincident with the experimental observations: K exhibited a peak at the stress-concentration position, and the maximum gradient K_{max} increased with increasing crack length. When the crack length increased from 0.1 to 1 mm, the normalized peak amplitudes K_{max} remarkably increased. However, with a further increase in the crack length from 2 to 3 mm, the growth ratio of K_{max} decreased. The trend of K_{max} is similar to the exponential variation in the crack propagation process, as shown in Fig. 6.9b. The relation between K_{max} and the length h can be effectively fitted by the given exponential fitting equation. However, an approximate linear relationship was reported between the maximum gradient K_{max} and crack length $2a$ in the crack propagation process throughout both the tension-tension test and bending fatigue test. These findings are not consistent with the test results. This could be because theoretical exponential fitting could be described as linear fitting when the crack length is smaller than a certain value, which is 1.2 mm, as shown in Fig. 6.9b. Moreover, the measured magnetic memory signals are inevitably disturbed by the crack width in the tests. The theoretical impact of different crack widths b and crack lengths h on the maximum gradient is shown in Fig. 6.10.

We propose that the maximum gradient K_{max} could be used to describe the stress concentration degree by the following Eq. (6.4).

$$\alpha = K_{max}/K_{std} \tag{6.4}$$

where K_{std} is the reference gradient value of magnetic signals for the ferromagnetic steel, e.g., the average gradient value in a scanning line or a given area, or K_{max}

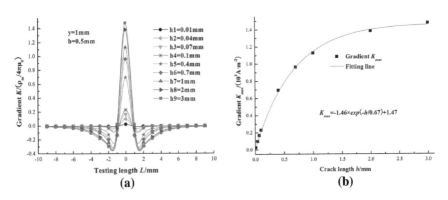

Fig. 6.9 Variation in the gradient K: **a** impact of the crack length on the normalized K amplitudes and **b** maximum gradient K_{max} and its fitting line

Fig. 6.10 Impact of the crack width b and crack length h on the variation in the maximum gradient K_{max}

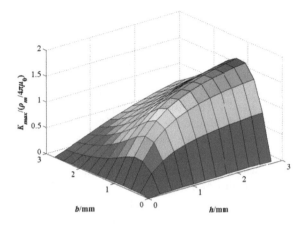

measured in a standard specimen at a given loading cycle. Thus, this simple formula could be used to quantitatively test the stress concentration degree of various areas in ferromagnetic material without identifying its stress state for engineering applications, although the method is still an approximation, and more work is needed to modify it.

6.2.3 Contributions of Stress and Discontinuity to MMM Signals

In fact, the variation in magnetic signals induced by the aforementioned stress concentration was essentially the result of the combined effect of stress and geometric discontinuity. To determine which is the major contribution to the abnormality of MMM signals, we used COMSOL Multiphysics simulation software to establish the model and analyze the magnetic signals. A specimen with a groove was placed in a sphere box filled with air, as shown in Fig. 6.11. The electromagnetic field distribution was calculated by Maxwell's magnetic equations. On the outside boundary of the

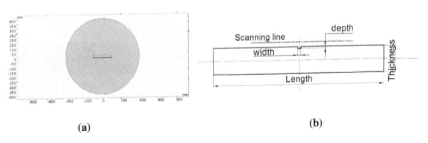

(a)   (b)

Fig. 6.11 Schematic of the finite element model: **a** specimen and surrounding air and **b** geometry dimension of the specimen with a groove

surrounding air box, the magnetic scalar potential could be considered zero because the diameter of the air box is large enough. On the air and specimen interface, the continuous boundary conditions are satisfied. The parameters used in the model can be seen in Ref. [13]. The initial magnetization of each mesh in the finite element (FE) model is assumed to be zero. After loading, the magnetization is determined by its stress, which is calculated from the magnetomechanical model as introduced in Chaps. 2 and 4. The magnetic fields generated by each mesh are merged to form the magnetic field around the specimen. Then, the MMM signals are extracted from the solution data along the scanning line, as shown in Fig. 6.11b.

The magnetic flux density B distribution caused by stress and geometric discontinuity with a tensile load of 20 kN is shown in Fig. 6.12 and Fig. 6.13, respectively. An abnormality of B can be observed near the groove. For the magnetic field induced by stress, the maximum magnetization area is located at the bottom of the groove, and the magnetization on the side of the groove is the weakest. This can be attributed to the free boundary condition releasing the stress concentration on the surface of the defect. However, the distribution of the magnetic field induced by geometric discontinuity is more complicated.

To better understand the impact of the geometric defect and stress distribution of the groove, the curves of the variation of the tangential signals H_x and normal signals H_y along the scanning line are plotted in Fig. 6.14. As expected, the H_x signal shows unipolar behavior, while the H_y signal shows bipolar behavior. The peak value of the H_x component corresponds to the location of the groove. Furthermore, the two types of curves have the same trend. An obvious decrease can be observed for the stress magnetization when compared to the geometric discontinuity, and the abnormality due to geometric discontinuity is even stronger than those due to stress. This indicates

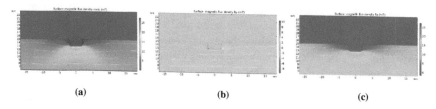

Fig. 6.12 Distribution of the magnetic field **a** norm B, **b** normal B_y and **c** tangential B_x caused by stress

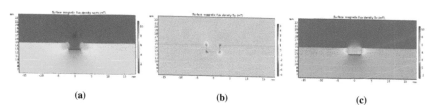

Fig. 6.13 Distribution of the magnetic field **a** norm B, **b** normal B_y and **c** tangential B_x caused by geometric discontinuity

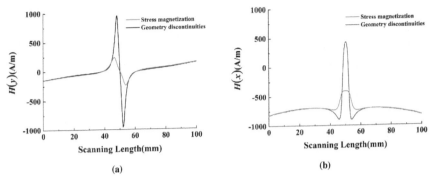

Fig. 6.14 Comparison of the magnetic signals due to the geometry discontinuity and stress magnetization: **a** normal signals H_y and **b** tangential signals H_x

that the geometry discontinuity is the main factor that determines the distribution of the MMM signals.

To further quantify the contributions of the stress and geometry discontinuity to MMM signals, two characteristic parameters are defined, namely, the peak-peak amplitude S_{p-p} and peak-peak width w_{p-p} of the normal component, as shown in Fig. 6.15. Then, the variation in S_{p-p} with increasing groove depth can be plotted in Fig. 6.16. The groove depth has a remarkable impact on the magnetic field amplitude S_{p-p} for both stress and geometric discontinuity. However, the value of S_{p-p} and its increment caused by geometry discontinuity are distinctly higher than those caused by stress. Furthermore, S_{p-p} decreases with increasing width under a constant depth under the condition of stress. However, it increases with increasing width under the condition of geometric discontinuity. This is because with the increase in the width, the slope of the normal component of MMM signal of the entire specimen weakens the characteristic parameters S_{p-p} under the condition of stress. However, under the condition of geometric discontinuity, because the value of S_{p-p} is much larger, the slope of the MMM signal of the entire specimen can be ignored, which has little effect on the variation in S_{p-p}. Figure 6.17 shows the w_{p-p} variation with increasing

Fig. 6.15 Curves and parameters of the normal component

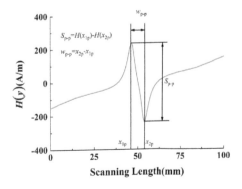

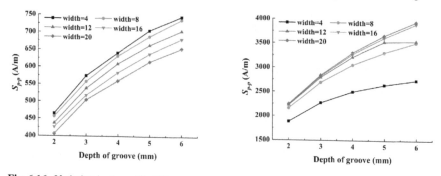

Fig. 6.16 Variation in S_{p-p} with different groove depths for **a** stress and **b** geometric discontinuity

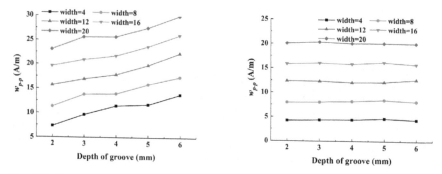

Fig. 6.17 Variation in w_{p-p} with different groove depths for **a** stress and **b** geometric discontinuity

groove depth. The groove depth has a slight influence on the peak-peak width w_{p-p} due to the stress, as shown in Fig. 6.17a. In contrast, the groove depth has little influence on the geometric discontinuity.

In the traditional magnetic flux leakage (MFL) method, magnetic charges are assumed to be concentrated in the walls of the breaking defect. Thus, w_{p-p} remains constant with changing depth. However, the stress is released in the walls of the groove, and the main reason for the magnetic leakage field is the magnetization caused by stress at the bottom of the groove. Since the measured signals come from the surface of the specimen, w_{p-p} can be seen as a far field characteristic of a defect in this case. Previous studies indicated that the peak-peak width increases with increasing lift-off value when the lift-off value is not very small, which explains why w_{p-p} increases with increasing depth. The results show that stress and geometric discontinuity can also affect the MMM signal and have a similar variation trend. In this study, maximum stress was observed at the bottom of the groove. Meanwhile, the free boundary condition releases the stress of the defect. This indicates that the geometry discontinuity is the main factor that determines the distribution of the magnetic field. The nonuniform magnetization induced by stress weakens the magnetic leakage field. It also reduces the peak-peak amplitude S_{p-p} and increases the peak-peak width w_{p-p}. The simulation results further explain the complex relation of the stress and geometry shape and promote the application of MMM.

6.3 Stress Concentration Evaluation Based on the Magnetic Dual-Dipole Model

6.3.1 Magnetic Scalar Potential

For geometric defects, such as rectangular gaps, triangular gaps, and closed cracks, the magnetic anomalies can be explained by the magnetic dipole model. Two primary criteria are commonly used to justify the defects: the normal component of the signal changes to its polarity, and the tangential component of the signal reaches a peak value. There is good agreement between the theory and experiment, as mentioned in Sect. 6.2. However, for the stress concentration caused by local plastic deformation, the waveform features are precisely opposite of the criteria. For example, Fig. 6.18 shows a specimen and its magnetic signals, which have been loaded in a three-point bending test. There is a stress concentration zone caused by local plastic deformation in the indentation and no notch on the surface. The magnetic signals were measured along the scanning line. In the test, the magnetic probe is a 3-axis anisotropic magneto resistance (AMR) sensor. Three components, H_x, H_y, and H_z, of the magnetic field

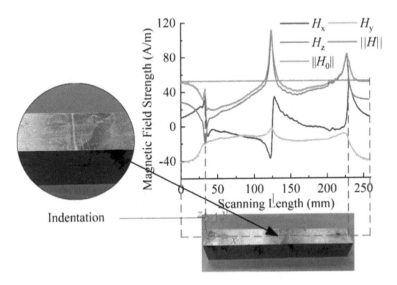

Fig. 6.18 Magnetic field signals of a specimen without a notch under a 15 kN compressive load. (||H|| is the amplitude of the magnetic field, and ||H_0|| is the amplitude of the environmental magnetic field measured before the test.)

can be measured, and the amplitude of the magnetic field $\|H\| = \sqrt{H_x^2 + H_y^2 + H_z^2}$ can be calculated, where $\|H_0\|$ is the amplitude of the environmental magnetic field measured before the test.

The tangential component of the signal (H_x and H_y) changes its polarity. The normal component (H_z) and the amplitude of the magnetic field ($\|H\|$) reach a peak value, while the measured environmental magnetic field $\|H_0\|$ remains constant. Several studies assume a linear distribution of stress concentration, but the signals calculated from this assumption may not match the waveform characteristics. Because there are no peak-to-valley characteristics, the stress concentration zone caused by local plastic deformation cannot be simplified as a dipole. Therefore, a dual-dipole model was proposed to evaluate the stress concentration caused by local plastic deformation.

From Maxwell's equations, which describe the electromagnetic field distribution on the macroscopic scale, for a ferromagnetic material, in the absence of outside currents, a static magnetic field can be described as:

$$\begin{cases} \nabla \times H = 0 \\ \nabla \cdot B = 0 \end{cases} \tag{6.5}$$

where H is the magnetic field strength and B is the magnetic flux density. The expression for the static magnetic field can be rewritten as follows by using the magnetic scalar potential (φ_m):

$$\begin{cases} \nabla^2 \varphi_m = 0, \quad (\text{in air}) \\ \nabla^2 \varphi_m = \nabla \cdot M, \quad (\text{in material}) \end{cases} \tag{6.6}$$

where φ_m is the magnetic scalar potential defined by $H = -\nabla \varphi_m$. This implies that $\nabla \cdot M$ is the source of the magnetic field, and we define the magnetic charge density as $\rho = -\nabla \cdot M$.

On the interface of two materials, continuous conditions are satisfied as follows:

$$\begin{cases} \varphi_{m1} = \varphi_{m2} \\ \frac{\partial \varphi_{m1}}{\partial n} - \frac{\partial \varphi_{m2}}{\partial n} = M \cdot n \end{cases} \tag{6.7}$$

where n is the outside unit vector normal to the surface and φ_{m1} and φ_{m2} are the potentials on the interface of the two materials. Since the scalar potential on the interface is equal, the distribution of the magnetic potential inside the material can be inferred by measuring the magnetic potential on the boundary of the air domain.

For $\nabla \times H = 0$ in the air domain, the magnetic field is a conservative harmonic field, which is irrotational everywhere. The integral of H and the potential φ_m only depends on the starting point A(x_a, y_a, z_a) and ending point B(x_b, y_b, z_b). It is independent from the path of integration. Therefore, the potential φ_m in the air domain is a function of position. It can be calculated as:

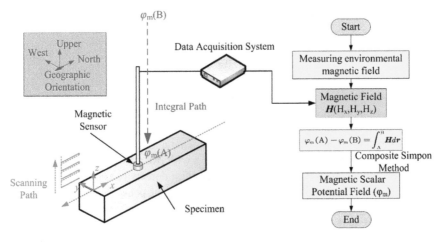

Fig. 6.19 Schematic of magnetic potential detection

$$\varphi_m(A) - \varphi_m(B) = \int_A^B \boldsymbol{H}\, d\boldsymbol{r} \qquad (6.8)$$

where $\boldsymbol{r}$ is the displacement vector $\boldsymbol{r} = x\boldsymbol{i} + y\boldsymbol{j} + z\boldsymbol{k}$ and i, j, and k are unit vectors along the x, y, and z directions, respectively.

To simplify the calculation, the vertical integral path is selected. As shown in Fig. 6.19, when the specimen is placed horizontally, the integration path **AB** is along the vertical direction (when $dx = 0$, $dy = 0$, and $dz =$ var). Then, φ_m can be calculated as:

$$\varphi_m(A) - \varphi_m(B) = \int_A^B \boldsymbol{H}\, d\boldsymbol{r} = \int_{z_a}^{z_b} H_z\, dz \qquad (6.9)$$

The approximation of this integration can be calculated by the composite Simpson method, a highly accurate numerical integration method. A schematic illustration of the calculation of the magnetic scalar potential is shown in Fig. 6.19. Before testing, the environmental magnetic field is measured, which can be regarded as a uniform field. It can be deducted as a DC component. Then, the magnetic scalar potential φ_m in the air domain can be obtained.

6.3.2 Magnetic Dipole and Its Scalar Potential

In general, the dipole model can effectively describe a simple defect at a distance of at least 3 times the size of a characteristic object. The magnetic scalar potential distribution around a dipole can be calculated as:

$$\varphi_m = -\frac{1}{4\pi} \boldsymbol{m} \cdot \frac{\boldsymbol{r}}{|\boldsymbol{r}|^3} \tag{6.10}$$

where $\boldsymbol{m} = \iiint_V \boldsymbol{M} dV$ and $\boldsymbol{M}$ is the material magnetization, and $\boldsymbol{r}$ is the distance vector from the dipole. Using this equation, the magnetic potential distribution of the dipole can be plotted, as shown in Fig. 6.20.

As can be observed, the magnetic potential generated by the dipole model has an obvious north and south (N-S) pole. The defect with two poles can be explained by this dipole model. However, for the specimen with defects caused by local plastic deformation, the distribution of the potential has only one peak, and the model should be modified as a dual-dipole model. The dual-dipole model assumes that there are two pairs of magnetic dipoles on the x-axis near the origin point, with a dipole moment of $\boldsymbol{m}$ and center distance of l, as shown in Fig. 6.21. When the center distance is sufficiently small, these two dipoles can be considered to be a dual dipole. Then, the magnetic potential generated by a dual dipole is:

$$\varphi_m = \frac{1}{4\pi} ml \frac{\partial^2}{\partial x^2}\left(\frac{1}{r}\right) = \frac{1}{4\pi} ml \frac{3x^2 - r^2}{r^5} \tag{6.11}$$

According to Eq. (6.11), the magnetic potential distribution of the dual dipole is plotted in Fig. 6.22. It has a maximum value at the position $x = 0$ and $y = 0$. This shows that the dual-dipole model can be used to represent the peak of the magnetic field distribution. To further study the effect of the dual-dipole angle, the variations in Hx, H_y, and H_z along different directions are calculated and shown in Fig. 6.22b–d. As can be observed, along the y-axis, the waveform features are almost consistent with the magnetic anomalies shown in Fig. 6.18.

Fig. 6.20 Distribution of the magnetic scalar potential for a magnetic dipole (lift-off = 1 mm, unit: normalization)

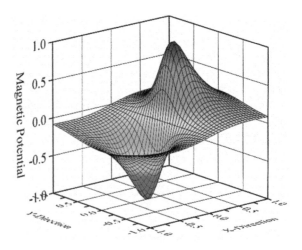

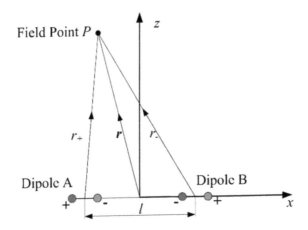

Fig. 6.21 Schematic diagram of magnetic dual dipoles

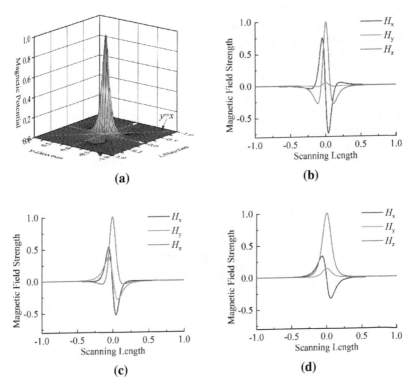

Fig. 6.22 The simulation results of the dual-dipole model: **a** The magnetic potential φ_m; and H_x, H_y, and H_z measured along the **b** x-axis direction; **c** $y = x$ direction; and **d** y-axis direction. (unit: normalization)

6.3.3 Measurement Process and Results

The specimen for verification experiments was made of Q235 steel whose magnetic parameters are shown in Table 6.2. Its microstructure is ferrite and a small amount of pearlite. Ferrite is a body-centered cubic form of iron and is a classic example of a ferromagnetic material. Its yield strength is 100–170 MPa. The pearlite is a two-phased, lamellar (or layered) structure composed of alternating layers of ferrite (88 wt%) and cementite (12 wt%). Cementite is a hard-brittle reinforcement in pearlite. It is ferromagnetic with a Curie temperature of approximately 230 °C. Although ferrite and pearlite are both anisotropic materials at the microscopic scale, the specimen can be considered isotropic at the macroscopic scale because of the random grain orientation at the macroscopic scale. Under the influence of mechanical loads, due to the negative mismatch in the thermal expansion coefficients between the ferrite and cementite phases, micro residual tensile stresses occur in the material. With increasing tensile loads, the magnetization process gradually shifts towards a magnetostrictive negative rotation.

The shape and dimensions of the specimen are shown in Fig. 6.23. This problem can be simplified to a 2D plane strain problem. This means that only σ_{xx}, σ_{yy}, and τ_{xy} should be considered, and the other components of the stress tensor are zero. The average demagnetization tensor of the specimen is $N = [N_x, N_y, N_z] = [0.0458,$

Table 6.2 Magnetic properties of the Q235 steel (SI units)

Material	Saturation magnetization M_s (MA/m)	Coercivity H_c (A/m)	Remanence B_r (T)	Maximum permeability μ_m
Q235	1.530	846	0.95	964

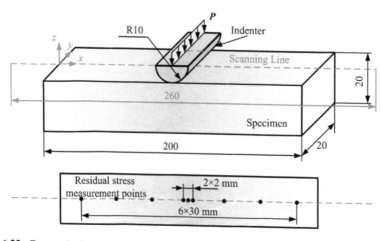

Fig. 6.23 Geometric dimensions of the specimen (unit: mm)

Fig. 6.24 Experimental setup for measuring the magnetic field

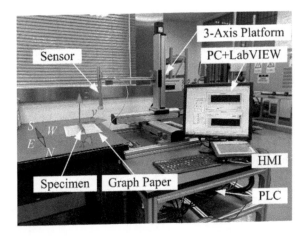

0.4771, 0.4771], where $N_x + N_y + N_z \approx 1.0$. A compressive load P is applied on an indenter with a radius of 10 mm by an SDS 100 testing machine. Before experiments, the specimens were demagnetized to eliminate the initial magnetic field caused by the initial stress. In testing, the specimen was first loaded to a predetermined level. Then, the loading machine was removed, and the magnetic signals were scanned by a 3-axis electronic platform with a given lift-off z (from 1 to 20 mm). Figure 6.24 illustrates the measurement setup. The precision of the ball screw in the platform was 0.018 mm/300 mm. The measuring range was 0.3 m $\times$ 0.3 m $\times$ 0.4 m, and the accuracy of the positioning of the platform was 0.02 mm. The scanning speed was 2 mm/s. The spatial resolution of the measurement was 0.28 mm. The measurement path was along the N-S direction, and the total length was 260 mm (longer than the total length of the specimen).

To reduce the positioning error of each repeat clamping, the specimen is fixed on a horizontal working plane. There are some locating points on the plane to ensure the fixed position of each measurement. The magnetic sensor is an HMC 5883 3D anisotropic magneto resistance (AMR) sensor. This is a digital integrated sensor. It integrates an amplifier and a 12-bit analog digital converter (ADC). The measured data are transferred to a personal computer (PC) by I2C Bus. The field range is ± 477 A/m, field resolution (noise floor) is 0.16 A/m, sensitivity (gain) is 2.89 LSB/A/m and hysteresis is ± 25 ppm. The residual deformation was measured by a Panasonic HG-C1100 laser displacement sensor with an accuracy of 70 μm and an effective measurement range of 100 $\pm$ 35 mm. The residual stress on the surface was measured by an X-ray stress analyzer (iXRD, Proto, Canada), referring to the standard ASTM E915-2010. The residual stress at each point was measured three times, and the mean value was recorded as the effective stress. The environmental magnetic field is a key factor for measurement. The three component values of the geomagnetic field in the laboratory are 26.2, -2.5, and 30.0 A/m, respectively, and the total field is 39.9 A/m. The geomagnetic field is calculated as the environmental magnetic field.

Figure 6.25 shows the distribution of the magnetic potential of the specimen in the xz plane under different loads. The potential method can effectively eliminate measurement errors. It improves the stability of the data and makes it easier to identify the magnetic source and geometric shape of the specimen. Figure 6.25a shows the distribution after demagnetization without loading. There is a clear formation of a magnetic dipole 200 mm in length. The position of the peak and valley (N-S poles) reflects the length of the specimen. The value of the N-S pole potential represents the magnetization of the specimen after loading. Figure 6.25b–f show the effect of the applied loads on the specimen. The polarity of the magnetic field reverses with increasing load. This is because the direction of the stress-induced magnetization is

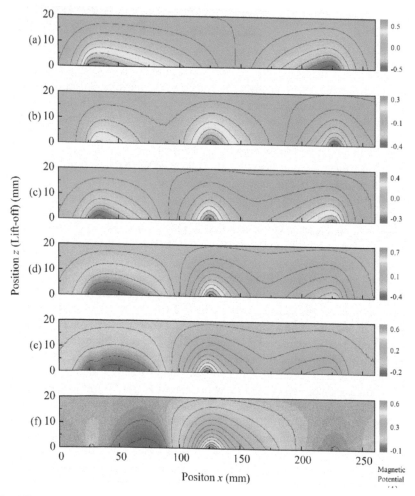

Fig. 6.25 The contour of the magnetic potential (unit: A) under applied loads: **a** 0 kN; **b** 5 kN; **c** 10 kN; **d** 15 kN; **e** 20 kN; and **f** 25 kN

opposite to the initial direction of the specimen. As the load increases, the stress-induced magnetization increases, causing the magnetic polarity to be reversed. In the indented area, there is a peak of magnetic potential at the loading area. The peak value increases with increasing loading. This indicates that the magnetization due to stress concentration is continuously enhanced. It can also be inferred that the peak of the magnetic potential can be used to characterize the stress concentration of the specimen. It should be noted that the peak value of the right pole is larger than the valley value of the left pole. The asymmetry of φ_m is caused by the asymmetry of the geometric boundaries, which will be discussed in the next section.

6.3.4   Analysis of the Magnetic Scalar Potential

The magnetization constitutive relation caused by the stress of the ferromagnetic material can be described by the magnetization curve. The J-A model is used to calculate the stress-dependent magnetization distribution, and the anhysteretic magnetization M_{an} can be obtained as introduced in Chap. 2:

$$M_{an} = M_s \left[\coth\left(\frac{H_{eff}}{a}\right) - \frac{a}{H_{eff}} \right] \tag{6.12}$$

Based on the above equations, the curve of Q235 steel in this study has been simulated with the reported parameter [14], as shown in Fig. 6.26. It can be seen clearly that the magnetization increases at first and then decreases. The curve can be modelled as a quadratic function:

$$M = f(\sigma) = A(\sigma - B)^2 + C \tag{6.13}$$

Fig. 6.26 Stress-dependent magnetization curve of 0.2 wt% carbon steel [14]

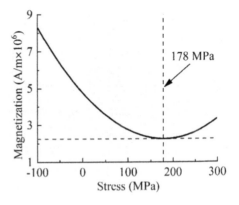

where A, B, and C are the parameters measured from the experiments: A = 7.664 × 10^{-8} Am/N, B = −178 MPa, and C = 2.2571 × 10^6 A/m. The coefficient of determination R^2 is 0.9848.

During the elastic stage, the magnetic charge inside the material can be calculated by:

$$\rho = -\nabla \cdot M = -\nabla \cdot f(\sigma) = f'(\sigma)\nabla \cdot \sigma \qquad (6.14)$$

For the material in the stage of plastic deformation, its density of dislocations considerably increases, and the magnetization induced by stress decreases sharply. As a result, there is a relatively low permeability in the contact zone. This shows that the divergence of the stress distribution ($\nabla\cdot\sigma$) is the determinant of the magnetic source. The stress distribution can be obtained from the Hertzian contact formula. As shown in Fig. 6.27, the stress is elliptically distributed on the contact line. The maximum stress is given by:

$$\sigma = \sqrt{\frac{FE^*}{\pi BR}} \qquad (6.15)$$

where F is the applied normal force, E^* is the combined elasticity modulus, B is the length of the cylinder (20 mm), and R is the combined radius.

The combined elasticity modulus is:

$$E^* = \frac{E_1 E_2}{E_2\left(1 - \mu_1^2\right) + E_1\left(1 - \mu_2^2\right)} \qquad (6.16)$$

Fig. 6.27 Schematic of the contact model

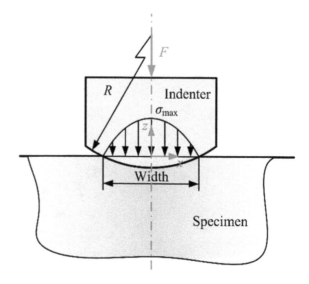

Table 6.3 The analytical contact stress on the specimen

Loading (kN)	Poisson's ratio $\mu_1 = 0.3$		Poisson's ratio $\mu_2 = 0.5$	
	Contact half width (mm)	Contact stress (MPa)	Contact half width (mm)	Contact stress (MPa)
5	0.168	946.7	0.160	992.5
10	0.238	1338.9	0.227	1403.5
15	0.291	1639.7	0.278	1718.9
20	0.336	1893.4	0.321	1984.9
25	0.376	2116.9	0.359	2219.2

The combined radius R is:

$$R = \lim_{R_2 \to \infty} \frac{R_1 R_2}{R_1 + R_2} \tag{6.17}$$

The contact width $2a$ is given by:

$$a = \sqrt{\frac{4FR}{\pi BE^*}} \tag{6.18}$$

Using the above equations, the contact width and maximum contact stress under different loadings can be calculated, as shown in Table 6.3. The Young's modulus and Poisson's ratio of the indenter are 210 GPa and 0.3, respectively. Based on the results, the maximum contact stress exceeds the elastic limit, and the material in the contact zone will enter the plastic deformation stage. In the plasticizing zone, Poisson's ratio changes from approximately 0.3–0.5, and the contact stress increases slightly.

To further study the distribution of the contact stress, a finite element (FE) analysis was performed using Comsol 5.3 software. The contour of the stress near the contact area under a typical load (15 kN) is shown in Fig. 6.28. As observed in Fig. 6.28a–c, the stress in the contact area exceeds the yield limit σ (235 MPa). In Fig. 6.28d, two peaks of the stress divergence are generated at both edges of the contact width. The left panel can be considered a magnetic dipole, as well as the right panel. Since the contact width is sufficiently small, these two dipoles can be regarded as a whole or a dual dipole.

As shown in Fig. 6.29, the two stress-induced magnetization peaks produce two pairs of dipole waveform features. The combination of these two dipoles yields the final variation profiles of the magnetic signals. To study the quantitative relationship between the density of the dual-dipole source and the applied load, the measurement data are further processed. Figure 6.30 shows the relationship between the peak value of the magnetic potential and macro stress. When the stress is less than 150 MPa, the peak value of the magnetic potential increases with increasing stress; otherwise, the peak value decreases gradually. It can be inferred that the inflection point of the

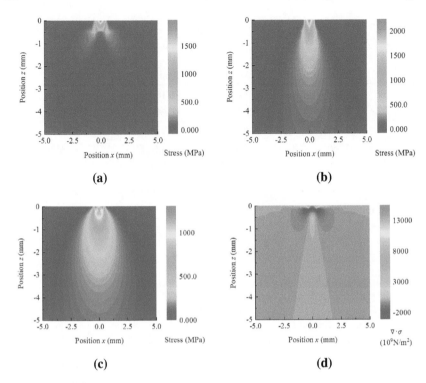

Fig. 6.28 Typical results of the stress distribution: **a** x-direction stress σ_x; **b** z-direction stress σ_z; **c** von Mises stress σ; and **d** divergence of the stress: $\nabla \cdot \sigma = \frac{\partial \sigma_x}{\partial x} + \frac{\partial \sigma_z}{\partial z}$

curves represents the Villari critical point (178 MPa for the material in the test). The error of the inflection point (150–178 MPa) is within the error range.

The residual deflection deformation of the specimen is shown in Table 6.4. When the load is greater than 15 kN, a substantial plastic residual deformation occurs. This causes the scanning line to not be parallel to the surface of the specimen. The variation in the lift-off value, as shown in Fig. 6.31, causes the magnetic potential at the N-S poles of the specimen to not be equal. The peak of the magnetic potential of the stress concentration zone decreases gradually.

To further investigate the variation in magnetic signals in the air domain, the magnetic potentials on the surface of the stress concentration zone along the z-direction were extracted from the contour map in Fig. 6.25b–f. Each specimen was measured 5 times. Figure 6.32 illustrates the magnetic potential with different lift-off values. This shows that the potential decreases exponentially with increasing lift-off. When the lift-off increases to 20 mm, the magnetic potential is close to zero. This means that the magnetic field anomalies caused by contact damage have nearly disappeared. It can be calculated that the signal–noise ratio (SNR) is larger than 10 dB when the lift-off is less than 11 mm. This means that the left area is the possible range of single magnetic anomaly detection based on the SNR criterion.

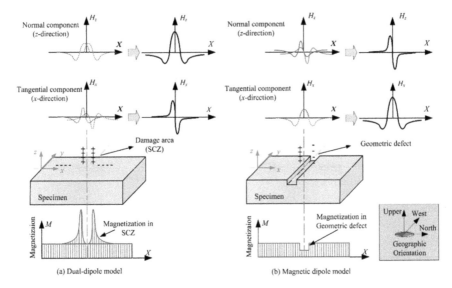

Fig. 6.29 A comparison of the magnetic signal characteristics between **a** the dual-dipole model for the stress concentration zone with contact damage and **b** the magnetic dipole model for a geometric defect

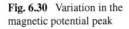

Fig. 6.30 Variation in the magnetic potential peak

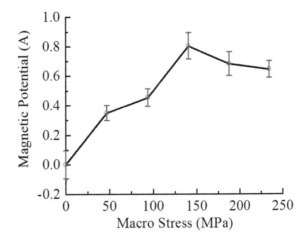

Table 6.5 shows the fitting parameters. In the fitting equation, $\varphi_m = A*\exp(-h/t) + \varphi_{m0}$, where A is related to the dipole moment $\boldsymbol{m}$ in the dual-dipole model, t stands for the decay rate in air, and φ_{m0} is related to the environmental magnetic field. The value of R^2 is very close to 1, representing the high accuracy of the fitting.

The residual stress along the x-direction is illustrated in Fig. 6.33. Before loading ($P = 0$ kN), the residual stress fluctuates between +15 and +45 MPa. This initial tensile stress may be caused by the milling process. After loading, the residual stress

Table 6.4 The residual deflection deformation of the specimen under loading

Loading (kN)	Maximum residual deflection deformation (mm)		
	No. 1	No. 2	No. 3
5	0.01	0.01	0.01
10	0.02	0.01	0.01
15	0.04	0.03	0.04
20	0.13	0.16	0.18
25	0.31	0.30	0.35

Fig. 6.31 The lift-off varies with the deformation of the specimen

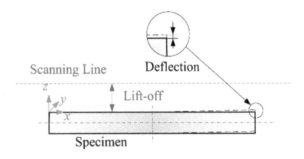

Fig. 6.32 Variation in the peak value of the magnetic potential with lift-off

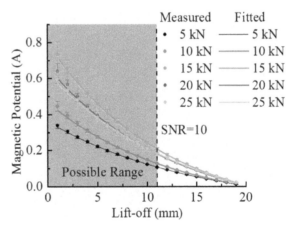

Table 6.5 The parameters of exponential fitting (φ_m: magnetic potential, and h: lift-off value)

Load (kN)	Exponential fitting equation: $\varphi_m = A*\exp(-h/t) + \varphi_{m0}$			
	A	t	φ_{m0}	R^2
5	0.48000 ± 0.00158	14.62282 ± 0.23113	-0.11883 ± 0.00300	0.99985
10	0.57631 ± 0.00171	12.22148 ± 0.19819	-0.10820 ± 0.00317	0.99987
15	0.95719 ± 0.00234	11.54336 ± 0.10360	-0.16359 ± 0.00260	0.99992
20	0.82075 ± 0.00236	12.40421 ± 0.16099	-0.15802 ± 0.00332	0.99992
25	0.81773 ± 0.00232	13.75914 ± 0.20131	-0.18561 ± 0.00426	0.99987

Fig. 6.33 The distribution
of the residual stress on the
surface of the specimen

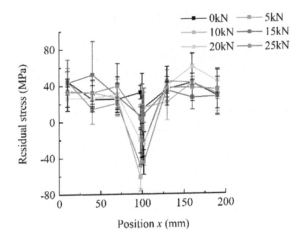

at the point far from the stress concentration zone ($x = 10, 40, 70, 130, 160$, and
190 mm) fluctuates between +20 and +60 MPa. This means that the applied loading
has essentially no effect on the residual stress. However, for the damaged area ($x = 98, 100$, and 102 mm), the residual stress changes substantially. The maximum
residual stress is −60 MPa, which is compressive stress. This means that local plastic
deformation changes the residual stresses from tensile stress to compressive stress.
It can also be inferred that at the edges of the contact area, two large stress variations,
from tension to compression, were generated. Each variation can be considered to
be the source of a dipole.

   In the experiments, when local plastic deformation occurs on the surface of the
material, the normal component of the MMM signals has a single peak at the center
of the stress concentration zone. The tangential component of the MMM signals
follows a peak-valley change through zero. This can be explained by the dual-
dipole model. This model can be used to evaluate the stress state of the material
before cracks and macro deformation appear, which has been confirmed by theoretical analysis and XRD residual stress tests. This shows that the model is viable
for predicting, assessing, and evaluating the stress concentration zone due to local
plastic deformation. In this study, the fluctuations in geometry on the surface caused
by plastic deformation were ignored. If the load increases, a nonnegligible geometric
notch will be formed due to the plastic flow of the material. Therefore, the magnetic
permeability of the notch zone declines sharply, which is far lower than that of the
undamaged ferromagnetic material. The magnetic anomalies caused by this kind of
defect can be explained by the traditional magnetic dipole model. When two or more
anomalies are close to each other, the observation range will change. Under this
circumstance, probes with higher accuracy and resolution are necessary to detect the
detailed information of the magnetic anomalies, which need to be further investigated.

6.4 Stress Concentration Inversion Method

6.4.1 Inversion Model of the Stress Concentration Based on the Magnetic Source Distribution

To quantitatively determine the stress concentration zone and its degree, the magnetic scalar potential and the magnetic source distribution should be further analyzed. The solution of the inversion of the stress concentration is calculated based on the iteration algorithm. And the whole calculation process can be described as follows.

First, the solution region is discretized as $N = m \times n$. The relationship between the magnetic source density parameter ρ_{ij}, the magnetic potential $G(\rho_{ij})$ and the testing point φ_{ij} can be represented as:

$$\varphi_{ij} = \sum_{i=1,j=1}^{mn} G\left(\rho_{ij}\right) \tag{6.19}$$

Thus, N element linear equations can be constructed when N linearly independent testing values are chosen. The matrix form can be written as $GX = V$, where G is the coefficient matrix, X is the magnetic charge density distribution vector, and V is the testing value vector of the magnetic potential distribution. The morbidity of the equations may be induced during calculation because small perturbations of input can lead to great changes in the solution. To overcome this problem, some algorithms, such as the conjugate gradient method and regularization method, can be used.

(1) Conjugate gradient method

The minimum error can be obtained based on the least square method, and thus, the inverse problem can be transformed to the minimum error:

$$\varepsilon = \min \| GX - V \|^2 \tag{6.20}$$

The initial estimation value x_0 is set up based on the system properties. The optimal solution of $GX = V$ is iteratively calculated:

$$x_{k+1} = x_k + \alpha_k d_k \tag{6.21}$$

where α_k is the iteration step length and d_k is the iteration direction. The ideal iteration method can reach the extreme point for specific accuracy through finite iteration steps. The inversion of the magnetic charge distribution can be realized based on the following function:

$$\min f(x) = \frac{1}{2} x^T A x + b^T x + c \tag{6.22}$$

where A is the positive definite symmetric matrix and b is the constant matrix. In addition, they satisfy the following relationship:

$$\begin{cases} A = 2G^T G \\ b = -2G^T V \end{cases} \tag{6.23}$$

The key is to solve the roots of equation $Ax + b = 0$. The global optimal solution can be calculated by constructing a set of conjugate directions.

(2) Regularization method

For an actual state electromagnetic field, the solution of the equation is unique. However, the noise component can also be found in the solution because the signal data collected by the sensor are easily affected by the external environment. Therefore, the regularization method is introduced in the solving process of the inverse problem. The oscillation item of the inverse problem is formed due to the integral operation when calculating the magnetic field based on the magnetic charge. The integral operator can relieve the effect of the high-frequency oscillation component. The morbidity of the equations can be modified by adding additional solution information, such as monotonicity, boundedness, an error range, etc. The Tikhonov regularization method is one of the most widely used regularization methods at present. The Tikhonov functional is constructed first:

$$J[X, V] = \|GX - V\|^2 + \alpha \|X\|^2 \tag{6.24}$$

where $\alpha > 0$ is the regularization parameter. The global minimal value X of functional $J[X, V]$ is the regularization solution of $GX = V$. Because the right side item of the equation for the inverse problem is generally obtained by testing, the suitable regularization parameter $\alpha*$ should be the root of the deviation equation when the error level is $\delta > 0$ based on the Morozov deviation principle method:

$$f(x) = \|GX - V\|^2 - \delta^2 \tag{6.25}$$

The solution can be calculated by the Newton iteration algorithm:

$$\alpha_{k+1} = \alpha_k - \frac{f(\alpha_k)}{f(\alpha_{k+1})} \tag{6.26}$$

where $f'(\alpha) = -2\alpha \left(\frac{dx_\alpha}{d\alpha}\right)^T x_\alpha$ and $\left(\frac{dx_\alpha}{d\alpha}\right)^T = -(G^T G + \alpha I)^{-1} G^T x$. α_0 will converge to the roots of the generalized deviation equation with the second-order rate when the suitable initial value $\alpha_0 > 0$ is set.

6.4.2   Inversion of a One-Dimensional Stress Concentration

For the plane stress concentration problem, when the size of the sample, as shown in Fig. 6.23, is much larger in one direction than that in the other two directions, it can be simplified as a one-dimensional stress concentration problem. First, the magnetic dual-dipole model is established as shown in Sect. 6.3. Then, the inversion of the stress concentration zone is conducted by testing magnetic signal data. The inversion results of the surface stress concentration of the sample are shown in Fig. 6.34. The inversion calculation processes include the direct method, the conjugate gradient method and the regularization method. It can be found that the direct method can lead to great oscillations. The conjugate gradient method can reflect the surface magnetic charge distribution of the sample, although the results also have a certain fluctuation. This is because both the amplitude and the frequency of oscillation can increase with an increasing number of iterations for the conjugate gradient method. The regularization method can recognize the magnetic charge distribution and the stress concentration zone very well. The inversion results of the stress concentration correspond to the experiments. This indicates that the regularization method can suppress the oscillation component during the inversion process.

To verify the inversion model and its method, the X-ray testing method was also conducted to measure the surface stress concentration of the sample. The testing points and their results are shown in Fig. 6.35. Before loading, the residual stress fluctuates from +15 to +45 MPa induced by the manufacturing process. After loading, the residual stress far away from the stress concentration zone fluctuates from +20 to +60 MPa, which is not affected by the applied load. However, the residual stress close to the stress concentration zone decreases sharply to −60 MPa. In general, the

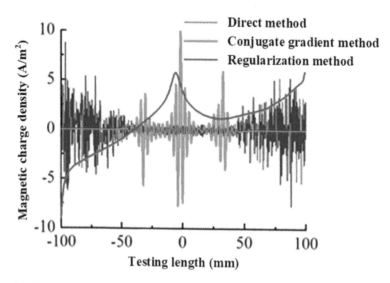

Fig. 6.34 The reversion results of the one-dimensional stress concentration

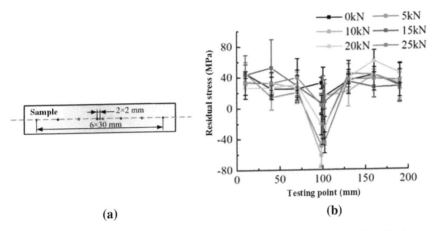

(a)

(b)

Fig. 6.35 Surface residual stress measurement of the sample **a** testing points; **b** residual stress distribution

regularization inversion method can determine the magnetic charge distribution of a sample, which can be used to recognize the stress concentration zone and degree.

6.4.3 Inversion of a Two-Dimensional Stress Concentration

To verify the feasibility of the inversion method for a two-dimensional stress distribution, the testing experiment was conducted in a plane sample whose shape and size are shown in Fig. 6.36. A load of 15 kN is applied at the center of the sample. A

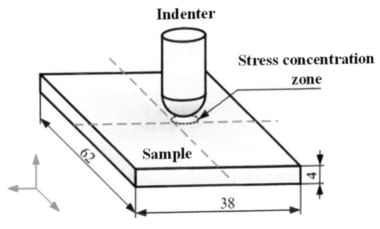

Fig. 6.36 Shape and size of the sample for the two-dimensional inversion experiment

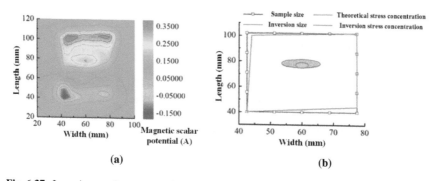

Fig. 6.37 Inversion results: **a** magnetic scalar potential distribution; **b** stress concentration zone

circular indentation with a diameter of 2 mm, which represents the stress concentration damage zone, is formed after unloading. The surface magnetic potential distribution of the sample is measured and plotted in Fig. 6.37a based on the magnetic dual-dipole model introduced in Sect. 6.3. Then, the inversion results of the stress concentration zone are shown in Fig. 6.37b based on the regularization inversion method. This indicates that this inversion algorithm can effectively determine the sample size and stress concentration zone.

6.5   Conclusions

The variation in the $H_p(y)$ signals of the specimens with different stress concentration factors was analyzed in tension-tension fatigue tests based on the magnetic dipole model. The impact of stress concentration on the variations in the MMM signals was studied. The $H_p(y)$ signals varied intensively and changed their polarity when cracks initiated at a loading cycle of 15,000 in the specimens with stress concentration factors of 4 and 5. The magnetic gradient of the $H_p(y)$ curve, K, was adopted to further describe the variation in the magnetic signals. Its maximum gradient, K_{max}, identifying the degree of stress concentration, was found to exponentially vary with the crack length. K_{max} is potentially a very useful indicator for monitoring stress concentrations in ferromagnetic structural steels. It is worth noting that the stress concentration with geometric discontinuity is the main reason for the variation in the MMM signals.

The magnetic anomalies of the stress concentration zone caused by local plastic deformation were also studied by proposing a new dual-dipole model. To reduce the measuring data errors, magnetic potential measurement and analysis were conducted, which is more convenient for determining the location of the magnetic anomalies. When the applied load acts on the surface of the specimen, an obvious peak of the magnetic potential is generated in the stress concentration zone, and the value of the peak can be used to qualitatively evaluate the stress concentration. The results showed that the magnetic field anomalies caused by local plastic deformation can be effectively represented by the dual-dipole model. This theoretical analysis was confirmed by the measured residual stress. With increasing load, the peak values of the magnetic potential first increase and then decrease. The variation was influenced by the applied loading and the deflection of the tested specimen. In the air domain, the magnetic potential of the dual-dipole model decayed exponentially with increasing lift-off values.

To finally quantitatively determine the stress concentration zone, the inversion model is introduced, and the regularization inversion algorithm is found to be the most effective method to suppress the oscillation problem during the inversion process. Based on the magnetic potential signal, the stress concentration evaluation can greatly decrease the effect of testing errors and improve the accuracy of quantitative analysis.

References

1. A. Fatemi, L. Yang, Cumulative fatigue damage and life prediction theories: a survey of the state of the art for homogeneous materials. Int. J. Fatigue **20**(1), 29–34 (1998)
2. C.L. Shi, S.Y. Dong, B.S. Xu et al., Stress concentration degree affects spontaneous magnetic signals of ferromagnetic steel under dynamic tension load. NDT&E Int. **43**(1), 8–12 (2010)
3. S. Bao, H. Lou, M. Fu et al., Correlation of stress concentration degree with residual magnetic field of ferromagnetic steel subjected to tensile stress. Nondestruct. Test. Eval. **32**(3), 255–268 (2017)
4. L.Q. Zhong, L.M. Li, X. Chen, Magnetic signals of stress concentration detected in different magnetic environment. Nondestruct. Test. Eval. **25**(2), 161–168 (2010)
5. L. Zhong, L. Li, X. Chen et al., Simulation of magnetic field abnormalities caused by stress concentrations. IEEE Trans. Magn. **49**(3), 1128–1134 (2013)
6. P.P. Shi, X.J. Zheng, Magnetic charge model for 3D MMM signals. Nondestruct. Test. Eval. **31**(1), 45–60 (2016)
7. P. Ramuhalli, L. Udpa, S.S. Udpa, Electromagnetic NDE signal inversion by function-approximation neural networks. IEEE Trans. Magn. **38**(6), 3633–3642 (2002)
8. A. Khodayari-Rostamabad, J.P. Reilly, N.K. Nikolova et al., Machine learning techniques for the analysis of magnetic flux leakage images in pipeline inspection. IEEE Trans. Magn. **45**(8), 3073–3084 (2009)
9. W. Han, J. Xu, P. Wang, G. Tian, Defect profile estimation from magnetic flux leakage signal via efficient managing particle swarm optimization. Sensors **14**(6), 10361–10380 (2014)
10. D.C. Jiles Theory of the magnetomechanical effect. J. Phys. D: Appl. Phys. **28**, 1537–1546 (1995)
11. M. Roskosz, M. Bieniek, Evaluation of residual stress in ferromagnetic steels based on residual magnetic field measurements. NDT&E Int. **45**, 55–62 (2012)
12. M. Roskosz, M. Bieniek, Analysis of the universality of the residual stress evaluation method based on residual magnetic field measurements. NDT&E Int. **54**, 63–68 (2013)
13. G. Han, H.H. Huang, Discussion of the influence of geometric discontinuity and stress concentration on the magnetic memory method. J. Magn. **25**(2), 269–276 (2020)
14. Y. Wang, X. Liu, B. Wu et al., Dipole modeling of stress-dependent magnetic flux leakage. NDT&E Int. **95**, 1–8 (2018)

Chapter 7
Temperature Impacts on MMM Signals

7.1 Introduction

When the magnet exceeds a certain temperature, its magnetism will be weakened due to the increase in temperature and the thermal movement of the atoms. The magnetic moment of the magnet atom tends to be arranged in a random state, which leads to a decrease in magnetization. Therefore, temperature is another important factor that can impact the magnetic field distribution of ferromagnetic materials apart from stress concentration.

At present, much attention has been given to determining how to evaluate the stress concentration and damage degree of ferromagnetic steels at room temperature for remanufacturing cores. However, a large number of equipment, such as steam turbines, steam pipelines, and boilers, are often operated under a high-temperature environment in the electric power industry. According to the operation standard of power stations in China, the maximum temperature of Q345 used in the steam outlet of medium- and low-pressure boilers is nearly 450 °C; the minimum temperature of high-pressure steam pipeline structural steel 20 g ranges from 200 to 300 °C; and the operating temperature of steam turbine final stage alloy steel blades approaches 50 °C. Therefore, the temperatures of these structural steel materials often remain between approximately 200 and 450 °C. Nevertheless, the change in the microstructures of ferromagnetic materials occurs at high temperatures, and steel exhibits a brittle fracture tendency after cooling. In addition, the surface quality of components is strongly degraded by the interaction of erosion and corrosion of water flow and heat steam. Therefore, creep deformation, corrosion, stress concentration, and fatigue damage easily occur in remanufacturing cores when ferromagnetic materials are subjected to complicated loads. If components are not diagnosed and evaluated in time, it is likely that the sudden failure of structures may occur and cause substantial losses. To identify the dangerous zones of ferromagnetic structural steel materials under a high-temperature environment, the variation in magnetic signals with temperature is also a challenging task that deserves to be further studied in remanufacturing.

© Science Press 2021
H. Huang et al., *Metal Magnetic Memory Technique and Its Applications in Remanufacturing*, https://doi.org/10.1007/978-981-16-1590-0_7

Ostash et al. [1] investigated the chemical and phase compositions, structural morphology, and mechanical properties of steel materials from steam pipelines of thermal power plants, and the results demonstrated that the coercive force H_c can be used to predict the residual service life of steam pipelines. Experience with carrying out comprehensive diagnostics of tube using the methods of metal magnetic memory (MMM), ultrasonic diagnostics, and metal structure analysis was described by Dubov [2]. It was shown that stress concentration zones caused by additional working loads were sources from which bending damage emerges and develops. Gladshtein [3] simulated the endurance of the metal of steam pipelines under conditions of creep, determined the relationship between the duration of different damage stages and the temperature, and found that the growth in the remaining life of the material was estimated based on a decrease in the temperature. Nevertheless, the experiments above only focus on the qualitative relation between the temperature and specific components, which cannot quantitatively illustrate the effect of temperature and stress on the magnetic field of ferromagnetic materials. Thus, it is necessary to discuss the effect of temperature variation on the surface magnetic field distribution of components when steam turbines, steam pipelines, and boilers are detected and maintained in the engineering field.

To investigate the mechanism of varied magnetic signals under a high-temperature environment, Ladjimi and Mekideche [4] developed a modified Jiles-Atherton (J-A) model to study the relationship between temperature and the magnetic hysteresis loop, but the applied loads were not considered. Therefore, in this chapter, a modified J-A model based on thermal and mechanical effects was first developed in Sect. 7.2. Then, the MMM signals were measured under different static tensile stress levels and temperature conditions in Sect. 7.3. A suitable magnetic signal characteristic value was proposed to study the effect of temperature on the distribution of the surface residual magnetic field in Sect. 7.4. In addition, the microstructures at the location of fracture under different temperature environments were observed, and the theoretical results from the modified model were also analyzed to verify the experimental results.

7.2 Modified J-A Model Based on Thermal and Mechanical Effects

According to the classical J-A model, the total magnetization M is the sum of the contributions of the irreversible M_{irr} and reversible M_{rev} magnetization components. In the model, M_{irr} is attributed to domain wall pinning, whereas M_{rev} is attributed to the reversible bowing of the domain walls. Hysteresis is generated because the nonmagnetic inclusions, grain boundaries, and internal stress block the magnetization process of domain wall pinning. The anhysteretic magnetization M_{an} curve of an isotropic material can be described with the modified Langevin function. Then, the expression of the hysteretic loop is determined based on the law of hysteresis loss and conservation of energy. To discuss the effect of temperature and stress on the intensity of the magnetic field, the classical J-A model is modified by introducing the thermal and mechanical effects.

7.2.1 Effect of Static Tensile Stress on the Magnetic Field

In some ways, an applied uniaxial stress acts like an applied magnetic field operating through magnetostriction. This additional field can be described by considering the energy A of the system along the reversible anhysteretic magnetization curve [5]

$$A = \mu_0 H M + \frac{\mu_0}{2}\alpha M^2 + \frac{3}{2}\sigma\lambda + T S \tag{7.1}$$

where $\mu_0\alpha M^2/2$ is the self-coupling energy, T is the temperature, and S is the entropy. The effective magnetic field causes a change in magnetization and therefore is determined by the derivatives of the energy A with respect to magnetization M

$$H_e = \frac{1}{\mu_0}\frac{dA}{dM} = H + \alpha M + \frac{3\sigma}{2\mu_0}\frac{d\lambda}{dM} \tag{7.2}$$

where $\frac{3\sigma}{2\mu_0}\frac{d\lambda}{dM}$ is the effective field component H_σ induced by stress, which means $H_\sigma = \frac{3\sigma}{2\mu_0}\frac{\partial\lambda}{\partial M}$. Based on the quadratic domain rotation model, the relationship between the magnetostriction coefficient λ of an isotropic material and the magnetization M can be described as [6, 7]

$$\lambda = \frac{3\lambda_S}{2M_S^2}M^2 \tag{7.3}$$

Then, Eq. (7.2) can be rewritten as

$$H_e = H + \left(\alpha + \frac{9\lambda_S\sigma}{2\mu_0 M_S^2}\right)M \tag{7.4}$$

where λ_S is the saturation magnetostriction coefficient, μ_0 is the permeability of a vacuum, M is the magnetization, M_S is the saturation magnetization, H is the geomagnetic field, and σ is the applied uniaxial stress.

For an isotropic material, the anhysteretic magnetization M_{an} is provided by the Langevin function in the J-A model [8, 9]

$$M_{an} = M_S\left[\coth\left(\frac{H + \alpha M_{an}}{a}\right) - \frac{a}{H + \alpha M_{an}}\right] \tag{7.5}$$

where α is the mean field parameter representing interdomain coupling and a characterizes the shape of the anhysteretic curve. The specimens were tested under static tensile loads without cyclic stress, so the hysteresis phenomenon is not presented, and the specimens reach the magnetic saturation state along the path of the initial magnetization curve, which can be considered to coincide with the anhysteretic magnetization curve, namely,

$$M = M_S\left[\coth\left(\frac{H + \alpha M}{a}\right) - \frac{a}{H + \alpha M}\right]$$

$$(7.6)$$

7.2.2 Effect of Temperature on the Magnetic Field

Many ferromagnetic structural steel materials have been subjected to the effect of a high-temperature environment in engineering applications, which changes the properties of the material; thus, the classical J-A model needs to be modified, and the temperature effect is introduced in the classical J-A model. The five parameters of M_S, a, α, k, and c need to be given in the classical J-A model; however, the parameters of k and c not needed in the calculation, and the variation in parameter a with temperature can also be ignored because of a small hysteresis loss [10]. Therefore, only the parameters of M_S and α are greatly affected by temperature.

The temperature dependence of saturation magnetization M_S can be expressed using the Weiss theory of ferromagnetism [11]

$$M_S(T) = M_S^{Ta}\left(1 - \exp\frac{T - T_c}{\tau_{M_S}}\right)$$

$$(7.7)$$

where M_S^{Ta} is the saturation magnetization at room temperature, T is the actual temperature, and T_c is the Curie temperature.

The mean field parameter α can be expressed as follows for an isotropic material [12]

$$\alpha = \frac{3a}{M_S} - \frac{1}{\chi'_{an}}$$

$$(7.8)$$

Anhysteretic susceptibility χ'_{an} is generally very high, so the second term of Eq. (7.8) is negligible compared with the first term, and the expression of the parameter α becomes

$$\alpha = \frac{3a}{M_S}$$

$$(7.9)$$

When the specimens are at room temperature

$$\alpha^{Ta} = \frac{3a}{M_S^{Ta}} \tag{7.10}$$

Substituting Eq. (7.7) into Eq. (7.9)

$$\alpha(T) = \frac{3a}{M_S^{Ta}} / \left(1 - \exp\frac{T - T_c}{\tau_{Ms}}\right) \tag{7.11}$$

Then, substituting Eq. (7.10) into Eq. (7.11)

$$\alpha(T) = \alpha^{Ta} / \left(1 - \exp\frac{T - T_c}{\tau_{Ms}}\right) \tag{7.12}$$

where τ_{Ms} is the relaxation time constant defined in the experiment and α^{Ta} is the mean field parameter at room temperature [13].

7.2.3 Variation in the Magnetic Field Intensity

The variation in the magnetic field intensity with static tensile stress and temperature can be given in Eq. (7.13), which is coupled by Eqs. (7.4), (7.6), (7.7), and (7.12). From the equation set, we can obtain the quantitative relation between the magnetic field intensity H_e and the tensile stress σ and temperature

$$\begin{cases} H_e = H + \left(\alpha + \dfrac{9\lambda_S\sigma}{2\mu_0 M_S^2}\right)M \\ M = M_S\left[\coth\left(\dfrac{H + \alpha M}{a}\right) - \dfrac{a}{H + \alpha M}\right] \\ M_S = M_S^{Ta}\left(1 - \exp\dfrac{T - T_c}{\tau_{Ms}}\right) \\ \alpha = \alpha^{Ta} / \left(1 - \exp\dfrac{T - T_c}{\tau_{Ms}}\right) \end{cases} \tag{7.13}$$

Table 7.1 Mechanical
properties of the specimens

Steel	Yield strength σ_s/MPa	Ultimate strength σ_b/MPa	Elongation rate/%
Q345	356	510	25.73

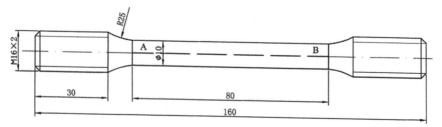

Fig. 7.1 Specimen shape and scanning line

7.3 Measurement of MMM Signals Under Different Temperatures

7.3.1 Material Preparation

The specimen was made of Q345 steel, which is a kind of structural steel widely used in boilers and steam pipelines because of its good impact toughness. Its chemical constitution and mechanical properties at room temperature are listed in Table 4.2 and Table 7.1. This steel material contains a few chemical elements of S and P, which can help to prevent brittle fracture. The shapes of six specimens marked from numbers 1 to 6 and their scanning lines are given in Fig. 7.1. Considering the normal operating conditions of structural steel materials in steam turbines, steam pipelines, and boilers, the temperature in the experiment was set from 25 to 250 °C. All the specimens were demagnetized before loading to eliminate the impact of the initial magnetic field.

7.3.2 Testing Method

The testing scheme can be divided into three different stages (i.e., the heating process, loading process, and detection process), as shown in Fig. 7.2. In the first stage, specimens 1–6 were heated for 5 min at 25 °C (i.e., room temperature), 50, 100, 150, 200, and 250 °C in a high–low temperature test chamber. Then, during the loading process, static tensile loads were applied to the specimens with an SDS-100 hydraulic testing machine, with stresses of 0, 102, 204, 306, and 408 MPa (i.e., loads of 0, 8, 16, 24, and 32 kN) until fracture. The specimens were unloaded and taken from the holders each time they were loaded to the preset stress. To protect the detection

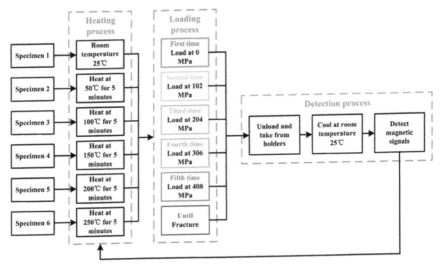

Fig. 7.2 Detection procedure of the magnetic signals

probe from the high-temperature environment and guarantee the detection precision, the specimens were cooled to room temperature. During the detection process, the specimens were laid on the platform in the south to north direction, the magnetic signals $H_p(y)$ above the scanning lines were measured by an EMS 2000+ MMM device at three different times, and the mean value of the measurements was taken as the final result to reduce the testing error of the magnetic signals $H_p(y)$. The probe with a 1 A/m sensitivity based on the Hall sensor was gripped on a nonferromagnetic 3-D electric scanning platform and was placed vertical to the surface of the specimen with a lift-off value of 1 mm and horizontal movement speed of 8 mm/s. After detection, the specimens were heated again to the preset temperature, and loading and detection were performed following the procedure until the specimens fractured.

The microstructures at the location of fracture that underwent a different high-temperature environment were observed by an MR5000 inverted metallurgic microscope at room temperature. Before examination, the specimens were sectioned, polished, and corroded with $FeCl_3$ saturated solution and Nital.

7.4 Variations in MMM Signals with Temperature and Stress

7.4.1 Normal Component of the Magnetic Signal

The variation in the normal component of the magnetic signal $H_p(y)$ under different temperature conditions with static tensile stress is shown in Fig. 7.3. Regardless of

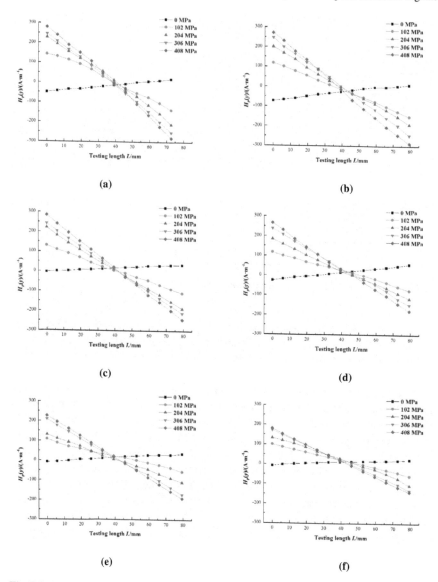

Fig. 7.3 Variation in the magnetic signals $H_p(y)$ under different temperature conditions: **a** 25 °C; **b** 50 °C; **c** 100 °C; **d** 150 °C; **e** 200 °C; and **f** 250 °C

how the temperature changed, all the magnetic signal $H_p(y)$ curves intersected at a distance of nearly 40 mm, where the magnetic signal value was approximately zero and presented good linearity along the scanning direction. The initial magnetic signals $H_p(y)$ remained stable with the value approaching zero because the specimens were demagnetized at the beginning and the direction of the magnetic field changed immediately when loaded at a small stress. The curves under different temperature

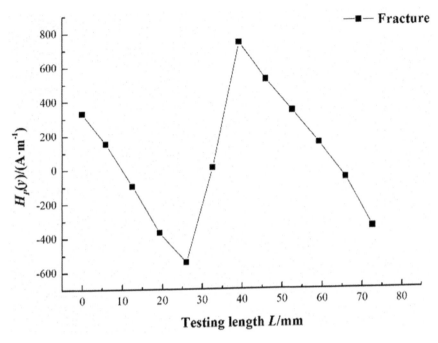

Fig. 7.4 Magnetic signals $H_p(y)$ at room temperature (25 °C) after fracturing

conditions rotated clockwise, which means that the intensity of the magnetic signal $H_p(y)$ and its absolute value of the gradient increased with increasing static tensile stress.

Specimen 1 fractured when loaded at 510 MPa (i.e., 40 kN) at room temperature (i.e., 25 °C), and the normal component of the magnetic signal $H_p(y)$ after fracturing is shown in Fig. 7.4. The $H_p(y)$ value changed dramatically from negative to positive, and the zero-crossing point of the curve appeared near the position of fracture.

7.4.2 Mean Value of the Normal Component of the Magnetic Signal

To better understand the variation in the magnetic signals with static tensile stress under different temperature conditions, the characteristic parameter $H_p(y)_{ave}$ was defined as the mean value of the normal component of the magnetic signal.

$$H_p(y)_{ave} = \frac{1}{n} \sum_{i=x_0}^{x_1} |H_p(y)_i| \qquad (7.14)$$

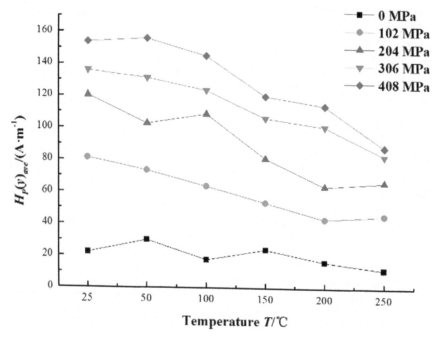

Fig. 7.5 Variation in the magnetic signals $H_p(y)_{ave}$ with temperature under different static tensile stress levels

where x_0 and x_1 are the starting point coordinate and the end point coordinate of the scanning line, respectively, and $|H_p(y)_i|$ is the absolute value of the normal component of the magnetic signal $H_p(y)$ of coordinate i.

Figure 7.5 shows the relation of $H_p(y)_{ave}$ with temperature under different static tensile stress levels. The magnetic signal $H_p(y)_{ave}$ gradually decreased with increasing temperature. In addition, the values of the magnetic signals $H_p(y)_{ave}$ at 250 °C under different tensile stress levels were reduced by at least 41% and up to 52% when compared with those at 25 °C. This means that the intensity of the magnetic signals at 250 °C was only approximately half of the value at room temperature (25 °C). Therefore, when equipment operated in a high-temperature environment is inspected and maintained, it should be noted that the intensity of the magnetic signals will seriously decrease. In addition, the magnetic signal $H_p(y)_{ave}$ increased with increasing static tensile stress at the same temperature, which indicated that a certain quantitative relationship may exist between the magnetic signal and the applied stress and temperature.

As shown in Fig. 7.6, the relationship between the magnetic signals $H_p(y)_{ave}$ and static tensile stress under different temperature conditions was fitted. The magnetic signals $H_p(y)_{ave}$ are similar to the exponential variation with the stress σ, which can be fitted by the following exponential fitting equation:

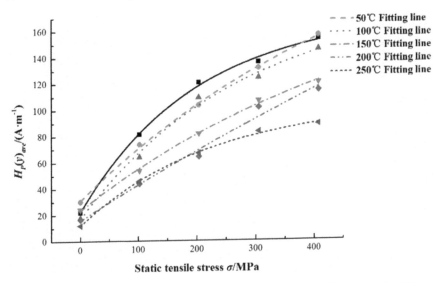

Fig. 7.6 Fitting relation of the magnetic signals $H_p(y)_{ave}$ with static tensile stress under different temperature conditions

$$H_p(y)_{ave} = \gamma + \beta \cdot \exp(\alpha \cdot \sigma) \tag{7.15}$$

where α, β, and γ are the coefficients related to temperature. The exponential relationship showed that the characteristic parameter $H_p(y)_{ave}$ can quantitatively reflect the degree of stress concentration. To discuss the rate of the change in the magnetic signals $H_p(y)_{ave}$ during the loading process of static tensile stress under the same temperature, the parameter of the gradient K can be further obtained by calculating the derivatives of the fitting function in Eq. (7.15):

$$K = \frac{d H_p(y)_{ave}}{d\sigma} = \alpha \cdot \beta \cdot \exp(\alpha \cdot \sigma) \tag{7.16}$$

In Eq. (7.15), all the fitting results show that $\alpha < 0$ and $\beta < 0$, so Eq. (7.16) is a decreasing function. The gradient K of the magnetic signals $H_p(y)_{ave}$ decreased with the increase in static tensile stress σ, which is shown in Fig. 7.7. The value of the gradient K approached zero when the specimens entered the plastic deformation stage. An explanation for this phenomenon is provided by the theory of the magnetic domain in ferromagnetic materials. During the elastic deformation stage, a large number of dislocations appeared, and the magnetic domains changed dramatically; thus, the intensity of the magnetic field increased quickly. When the specimens were very close to the plastic deformation stage, the dislocation density increased slowly with further tension. In the next moment, the dislocation walls piled up at the vicinity of the grain boundaries and blocked the movement of the domain walls, which led to the intensity of the magnetic field approaching the saturation state.

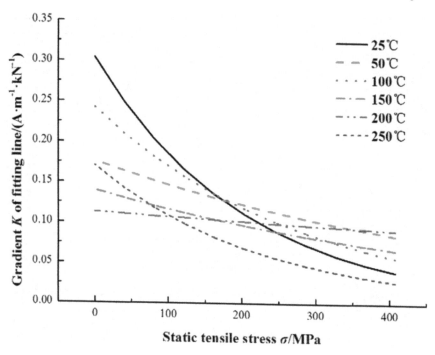

Fig. 7.7 Variation in the gradient K of the fitting line with static tensile stress under different temperature conditions

7.4.3   Variation Mechanism of the Magnetic Signals Under Different Temperatures

To understand how the microstructures under different temperature conditions relate to the variation in the magnetic signals, crystallographic textures at the location of fracture were observed, as shown in Fig. 7.8. The material of Q345 steel containing chemical elements of Si, P, and S can produce many inclusions of sulfide, silicate, and spherical oxide with increasing temperature because of the oxidation reaction. Figure 7.8a indicates that a uniform distribution of crystallographic textures with small black inclusions appeared at room temperature (25 °C). With increasing temperature, the concentration degree of the inclusions became increasingly obvious, as shown in Fig. 7.8b–f. These inclusions disturbed the uniform distribution of crystallographic texture and directly affected the quality and performance of ferromagnetic materials. The oxide inclusions with higher hardness and worse plasticity destroy the continuity of the base material; the silicate inclusions can easily produce plastic deformation with increasing temperature; and the coefficient of thermal expansion of sulfide inclusions is very different from that of the base material. Thus, the inclusions destroy the formation of metallurgical bonds with the base material, block the movement of dislocations in slip bands, and pin the grain boundary

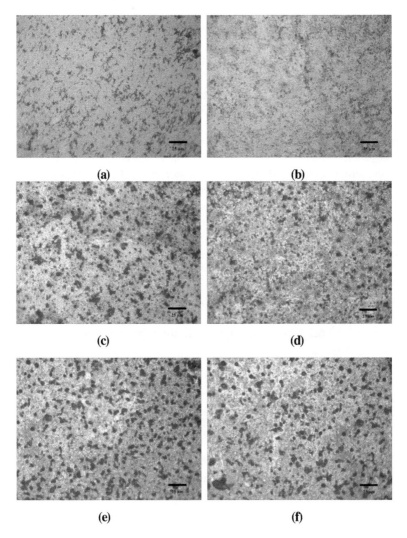

Fig. 7.8 Microstructures at the location of fracture under different temperature conditions: **a** 25 °C; **b** 50 °C; **c** 100 °C; **d** 150 °C; **e** 200 °C; and **f** 250 °C

migration of the metallographic structure. The magnetic signals are generated by the irreversible orientation of the magnetic domains and the accumulation of dislocations in ferromagnetic materials. However, the size of the inclusions increases with increasing temperature, which blocks the movement of the magnetic domains. As a result, a decrease in the magnetic signals is induced, and the results of the metallographic test are consistent with the variation in the magnetic signals $H_p(y)_{ave}$ under different temperature conditions.

7.4.4 Analysis Based on the Proposed Theoretical Model

The variation in the effective field H_e with static tensile stress σ and temperature T can be calculated based on Eq. (7.13), as shown in Fig. 7.9. According to the experimental environment and sample materials, the values of the model parameters were $\lambda_S = 30 \times 10^{-6}$, $\mu_0 = 4\pi \times 10^{-7}$, $M_S^{Ta} = 1.6 \times 10^6$ A m^{-1}, $H = 40$ A m^{-1}, $\alpha^{Ta} = 0.001$, $a = 1000$ A m^{-1}, $T_c = 770$ °C, and $\tau_{Ms} = 150$ s.

Figure 7.9 shows the relation of the effective field H_e with changing temperature T under different static tensile stress levels. The effective field H_e decreased with increasing temperature T and increased with increasing stress σ. In addition, if the stress σ was smaller, the effect of the temperature variation on the effective field H_e was weaker. The effective field H_e presented good linearity along the horizontal direction regardless of how the temperature changed when the value of stress σ was zero. To better understand the relationship among the static tensile stress σ, temperature T, and effective field H_e, the 3-D curved surface is shown in Fig. 7.10. The figure clearly shows that the effective field H_e decreased with the increase in temperature T and increased with the increase in static tensile stress σ.

The theoretical analysis agrees with that of the experimental results. This shows that the effective field H_e in the modified J-A model and the mean value of the normal component of the magnetic signal $H_p(y)_{ave}$ in the experiment can both effectively

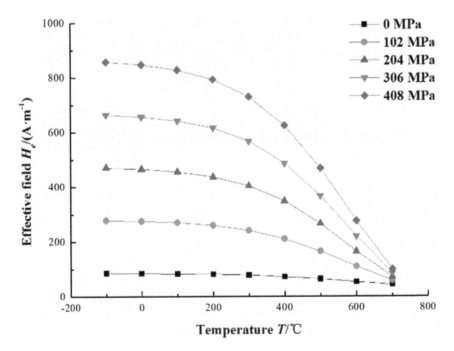

Fig. 7.9 Relation of the effective field H_e with changing temperature T under different static tensile stress levels

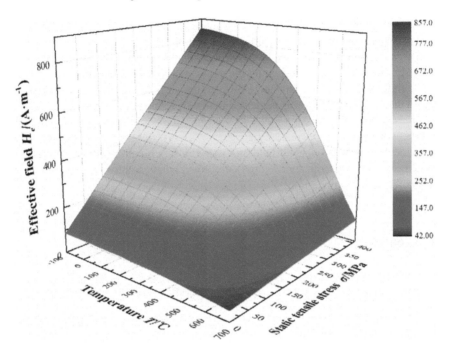

Fig. 7.10   3-D curved surface of the variation of the effective field H_e with changing static tensile stress σ and temperature T

characterize the magnetic field intensity and reflect the variation in the magnetic memory signals.

7.5   Conclusions

A quantitative relationship among the temperature, applied stress, and spontaneous magnetic signals in ferromagnetic steel materials was proposed, and the mechanism of varied magnetic signals under different stress levels and temperature conditions was established. In the experiment, ferromagnetic steel Q345 was tested; the magnetic memory signal $H_p(y)$ on the surface of the specimens increased with increasing applied stress. The mean value of magnetic signal $H_p(y)_{ave}$ decreased with increasing temperature but increased exponentially with increasing tensile stress; and the microstructures at the location of fracture under different temperature conditions showed that the size of the inclusions of ferromagnetic materials increased with increasing temperature. This blocked the irreversible movement of the magnetic domains, and thus, the intensity of the magnetic signals became weaker. A modified J-A model based on thermal and mechanical effects was proposed, and the analytic results indicated that the intensity of the effective field H_e decreased with increasing temperature T and increased with increasing static tensile stress σ, which showed that the experimental results were consistent with the theoretical analysis.

The experimental results and the theoretical analysis covered the shortcomings of nondestructive testing and evaluating the stress status of the remanufacturing cores in a high-temperature environment. However, the variation in the magnetic signals during the cooling process was neglected in the experiment, the effects of the phase transformation on the stress evolution were not considered in the modified J-A model, and the weak residual magnetic signals are easily disturbed by many factors, such as the environmental magnetic field and speed of heating and cooling. Therefore, more research needs to be done to apply this novel nondestructive testing method in remanufacturing engineering.

References

1. O.P. Ostash, O.V. Vol'demarov, P.V. Hladysh, Diagnostics of the structural-mechanical state of steels of steam pipelines by the coercimetric method and prediction of their service life. Mater. Sci. **49**(5), 667–680 (2014)
2. A.A. Dubov, Comprehensive diagnostics of the bends of boiler and steam-line tubes. Therm. Eng. **54**(9), 712–715 (2007)
3. V.I. Gladshtein, Estimation of the effect of stresses and temperature on damage accumulation in steam pipe bends by simulating the endurance of the metal by testing notched specimens. Met. Sci. Heat Treat. **53**(11–12), 611–617 (2012)
4. A. Ladjimi, M.R. Mekideche, Modeling of thermal effects on magnetic hysteresis using the Jiles-Atherton model. Przegląd Elektrotechniczny **88**(4a), 253–256 (2012)
5. D.C. Jiles, Theory of the magnetomechanical effect. J. Phys. D Appl. Phys. **28**(8), 1537 (1995)
6. M.J. Sablik, D.C. Jiles, Coupled magnetoelastic theory of magnetic and magnetostrictive hysteresis. IEEE Trans. Magn. **29**(4), 2113–2123 (1993)
7. S.Y. Cao, B.W. Wang, R.G. Yan et al., Dynamic model with hysteretic nonlinearity for giant magnetostrictive actuator. Proc. CSEE **11** (2003)
8. D.C. Jiles, Frequency dependence of hysteresis curves in conducting magnetic materials. J. Appl. Phys. **76**(10), 5849–5855 (1994)
9. Z. Li, Q.M. Li, C.Y. Li et al., Queries on the JA modeling theory of the magnetization process in ferromagnets and proposed correction method. Proc. CSEE **31**(03), 124–131 (2011)
10. P.R. Wilson, J.N. Ross, A.D. Brown, Simulation of magnetic component models in electric circuits including dynamic thermal effects. IEEE Trans. Power Electron. **17**(1), 55–65 (2002)
11. H. Bleuvin, *Analyse par la méthode des éléments finis des phénomènes magnéto-thermiques-application aux systèmes de chauffage par induction* (Grenoble Institute of Technology, Grenoble, 1984)
12. D.C. Jiles, J.B. Thoelke, M.K. Devine, Numerical determination of hysteresis parameters for the modeling of magnetic properties using the theory of ferromagnetic hysteresis. IEEE Trans. Magn. **28**(1), 27–35 (1992)
13. A. Raghunathan, Y. Melikhov, J.E. Snyder et al., Modeling the temperature dependence of hysteresis based on Jiles-Atherton theory. IEEE Trans. Magn. **45**(10), 3954–3957 (2009)

Chapter 8
Applied Magnetic Field Strengthens MMM Signals

8.1 Introduction

The research results in Chaps. 4–6 indicate that the metal magnetic memory (MMM) signals are essentially induced by stress concentrations. The relevant physical mechanism of the MMM signal has been studied, and various theories and models have been proposed. Some have found that the $H_p(y)$ signals are distributed randomly for different materials, especially in the elastic deformation stage, while others have found that the $H_p(y)$ curves have regular linear changes with increasing stress and have different features during the elastic and plastic deformation stages [1–3]. The reason for these different conclusions is that the MMM signal is essentially a type of weak spontaneous magnetic flux leakage field, which is measured under the excitation of a weak geomagnetic field. Therefore, MMM test results are easily affected by external conditions, such as the geomagnetic field, lift-off value of the probe, temperature, and environmental stray magnetic field. In particular, Chap. 7 discusses the decrease in the MMM signal under high temperatures in detail.

This means that the MMM signal is influenced by the environmental magnetic field during detection, and the test results may vary with changes in the external magnetic environment [4, 5], which easily causes unreliable detections such as missed detection and misjudgment. Currently, MMM signals are stimulated by geomagnetic fields, and their signal strengths are weak, most of which are within the range of ±100 A/m.

It is worth noting that magnetic flux leakage (MFL) detection can capture the higher leakage magnetic field of the defective parts under the action of a strong magnetic field. Therefore, inspired by MFL detection, we hope to enhance the outside magnetic excitation, thus improving the strength of the MMM signal and the signal-to-noise ratio. In this research, the MMM signals will be measured after the ferromagnetic materials are motivated by an external magnetic field. This means that the applied excitation magnetic field will be removed before MMM detection is conducted, which makes the proposed detection method different from that in MFL detection.

© Science Press 2021
H. Huang et al., *Metal Magnetic Memory Technique and Its Applications in Remanufacturing*, https://doi.org/10.1007/978-981-16-1590-0_8

Some studies try to analyze the influence of an external magnetic field on MMM detection. For example, Wang et al. analyzed the effect of a background magnetic field on the strength of the residual magnetic field and then proposed a corresponding method to eliminate this effect through finite element analysis [6]. As is well known, the geomagnetic field plays the role of an exciting source in generating a residual magnetic field, whose results may be interfered with by stray magnetic fields. The use of an applied magnetic field may strengthen the features of magnetic signals by eliminating the interference coming from stray magnetic fields. However, it is still unknown how the applied magnetic field affects the residual magnetic signals and how the magnetic signals vary with the applied magnetic field and stress.

In this chapter, tension-tension fatigue tests are carried out first on low-carbon steel remanufacturing cores when different magnetic field intensities are applied, as shown in Sect. 8.2. The quantitative relationships between the surface residual MMM signals and the applied magnetic field and cyclic stress are investigated in detail. In addition, static tensile tests are also carried out in Sect. 8.3 to determine the impact of the applied magnetic field and static tensile stress on the variation in the residual magnetic signals in ferromagnetic materials.

8.2 MMM Signal Strengthening Effect Under Fatigue Stress

The specimen is made of Q345 low-carbon steel whose yield strength is 358 MPa and ultimate strength is 484 MPa. The shapes of the sheet specimens are given in Fig. 8.1. Test lines were marked on each specimen, and these specimens were polished and demagnetized before loading.

An experimental system was established where which both the applied magnetic field and stress are controllable, as shown in Fig. 8.2. Dynamic tension loads were applied to the specimens on an MTS810 servo hydraulic testing machine, whose dynamic load error was within ±1.0%. Tension-tension fatigue tests with a constant amplitude (sinusoidal waveform) were performed, with the maximum stress at 333 MPa and the minimum stress at 33.3 MPa. The one-dimensional Helmholtz coil was placed vertical to the ground so that the direction of the applied magnetic

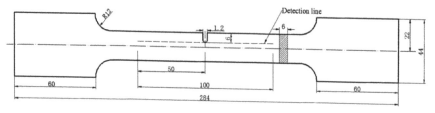

Fig. 8.1 Specimen shape and detection lines (in mm)

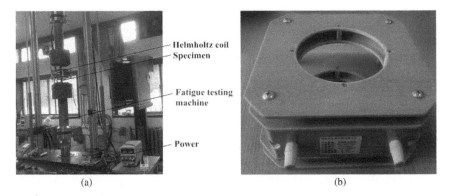

Fig. 8.2 Experimental devices: **a** the fatigue testing system; and **b** the one-dimensional Helmholtz coil

field was parallel to the tensile stress. Thus, the applied magnetic field worked on the specimen along with the geomagnetic field and cyclic tensile stress.

The experiments were carried out at room temperature. When the specimens were subjected to applied loads, the one-dimensional Helmholtz coil generated a constant magnetic field range from -1600 to 1600 A/m in the center of the coil. The applied magnetic field intensity can be precisely controlled by adjusting the current passing through the coil. An EMS2003 magnetic memory/eddy current detection diagnostic instrument was used for the measurements of magnetic memory signals offline. A two-channel probe was installed on a nonferromagnetic scanning platform, with a lift-off value of 0.5 mm and movement speed of 8 mm/s along the scanning lines during testing. The instrument was calibrated in the magnetic field of the Earth, with an assumed value of 40 A/m.

In the fatigue tests, the specimens were unloaded when the number of loading cycles reached 2000, 3000, 4000 and 5000, and they were examined away from the testing machine by laying them on the scanning platform in the south to north direction. For specimen 1, the current passing through the Helmholtz coil was set as the fundamental current $I_B = 0.24$ A. It produced a magnetic field whose intensity was $H_B = 200$ A/m. Adjusting the current passing through the Helmholtz coil to $1.5I_B$, $2I_B$ and $2.5I_B$ for specimens 2, 3 and 4 produced a series of magnetic fields with intensities of $1.5H_B$, $2H_B$, and $2.5H_B$, respectively.

8.2.1 Variations in the MMM Signals with an Applied Magnetic Field

To further describe the variation in the residual magnetic field, the gradient of the $H_p(y)$ curves, K, was given as follows:

$$K = \left| \Delta H_p(y) \middle/ \Delta L \right| \qquad (8.1)$$

where $\Delta H_p(y)$ is the differential value of the magnetic signals between two points and ΔL is the distance between the two points. The maximum gradient of $H_p(y)$ on the scanning line, K_{max}, was given as follows:

$$K_{max} = max\left(\left| \Delta H_p(y) \middle/ \Delta L \right| \right) \qquad (8.2)$$

Figure 8.3 shows the variation in the residual magnetic field. It can be seen that $H_p(y)$ is stable with the application of cyclic stress, and it increases with the increase in the applied magnetic field intensity H. With increasing stress cycles, the stress concentration becomes increasingly intensive at the location of the notch, where $H_p(y)$ changes sharply, crossing the zero value with a maximum gradient value K_{max}. Nevertheless, the $H_p(y)$ curves have good linearity at the location away from the stress concentration area, and the gradient K is constant and nearly unaffected by the stress cycles. It can be seen that K is only relevant to the applied magnetic field intensity H, as shown in Fig. 8.4. The gradient is approximately linear to the

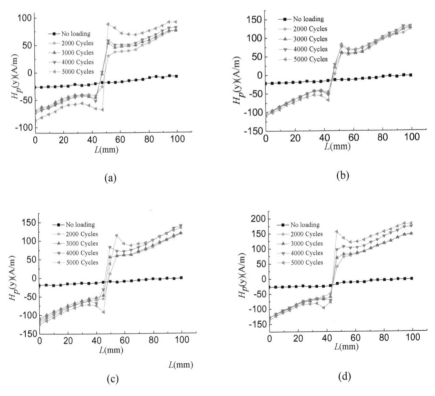

Fig. 8.3 $H_p(y)$ components of the residual magnetic field from measured lines excited by the applied magnetic field with different magnetic field intensities: **a** H_B; **b** $1.5H_B$; **c** $2H_B$; and **d** $2.5H_B$

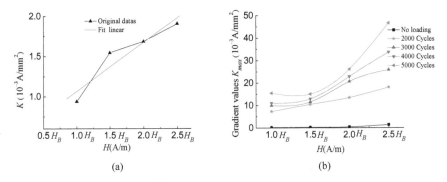

Fig. 8.4 The relation between the applied magnetic field intensity H and **a** the slope coefficient K of the $H_p(y)$ signal curve and **b** K_{max} appearing in the stress concentration area

magnetic field intensity, which is described by

$$K = aH + b \tag{8.3}$$

where a and b are both constant values related to the applied magnetic field, the material and the shape of the specimen. In the experiment, a is 0.61, b is 0.45. The gradient K and magnetic field intensity H are well fitted by Eq. (8.3) in Fig. 8.4a. In the fatigue tests, the applied maximum stress was very close to the yield limit of the Q345 material; thus, plastic deformation appeared immediately under the effect of fatigue loading. The specimen enters the plastic deformation stage and exhibits many dislocations after a few stress cycles, which prevents the domain wall from moving and weakens the magnetization [7]. Relaxation of internal stress occurs due to dislocation slip so that the magnetic signals are independent of the loading cycles. These findings are consistent with the observations in both cyclic tensile and bending tests [8–10]. Figure 8.4b indicates that K_{max} rises with increasing stress cycles or applied magnetic field intensities. This shows that the point of intersection of the $H_p(y)$ curves approximately crosses the zero value and recognizes the crack center, which is the most serious stress concentration zone. K_{max} increases with increasing stress cycles since the stress concentration at the zone approaching the notch of the specimen becomes increasingly intensive. Accordingly, K_{max} can be detected to identify the stress concentration in the specimen.

8.2.2 Theoretical Explanation Based on the Magnetic Dipole Model

Due to the existence of pre-notch on the specimen, the opposite magnetic charges will accumulate on the either side of the pre-notch. It is suitable to use the magnetic charge model which has been introduced in Chaps. 2 and 6 to analyze the variations of magnetic signals under the effect of applied magnetic field. The normal component $H_p(y)$ of the surface magnetic leakage field and its gradient K at the point of $P(x, y)$ can be expressed based on Eqs. (6.1), (6.2) and (6.3). And the profiles of $H_p(y)$ and K are shown in Fig. 8.5. Near the notch, $H_p(y)$ changes negatively-positively and has a zero value; at the same time, an abnormal gradient K of the magnetic memory signal can be observed, which explains the results of the fatigue tests shown in Fig. 8.3. This shows that the applied magnetic field only influences the distribution of the magnetic memory signals, including the maximum value and shapes. However, it does not affect the zero value position of the $H_p(y)$ curve.

Under the excitation of an applied magnetic field and cyclic stress, the magnetic charge density ρ_m increases, and both $H_p(y)$ and K are linear with ρ_m, which increases with increasing ρ_m, as shown in Eqs. (6.2) and (6.3). The variation in the magnetic charge density ρ_m at the location of the notch is shown in Fig. 8.6. This indicates that magnetic charges gather gradually in the stress concentration area under the action of an applied magnetic field and cyclic tensile stress. ρ_m increases with increasing applied magnetic field intensities or stress cycles, which produces higher $H_p(y)$ values, K and K_{max}.

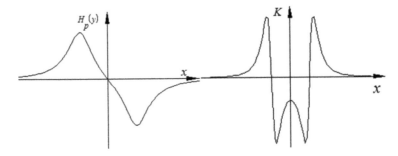

Fig. 8.5 The distribution of $H_p(y)$ and its gradient K

Fig. 8.6 Variation in the magnetic charge density ρ_m at the location of the notch

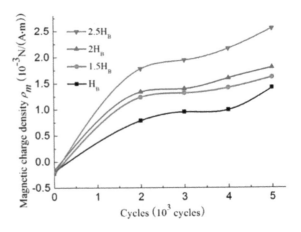

8.3 MMM Signal Strengthening Effect Under Static Stress

The static tensile stress is a type of particular case of cyclic stress when the fatigue number is zero. To study the effect of the applied magnetic field on the MMM signals under static tensile stress, static tensile tests were also carried out on carbon steel materials under different magnetic field strengths. The shape of the sheet specimens is given in Fig. 8.7. The specimens were labelled from No. 1 to 5, and five testing lines with lengths of 100 mm were marked on each specimen. Before loading, all the specimens were polished and demagnetized by a TC-2 demagnetizer, which is an AC induction demagnetization device.

The experimental system in Fig. 8.2 was also used in static tensile tests in which both the applied magnetic field and tensile stress are controllable. Static tensile loads with a loading rate of 20 kN/min were applied to the specimens on an electrohydraulic servo fatigue testing machine, whose load error was within $\pm 0.5\%$. The one-dimensional Helmholtz coil was placed vertical to the ground so that the direction of the applied magnetic field was parallel to the tensile stress. Thus, the applied magnetic field acted on the specimen along with the geomagnetic field and tensile stress. The one-dimensional Helmholtz coil (with a diameter of 90 mm and a length of 120 mm) can generate a constant magnetic field ranging from -1600 to 1600 A/m in the center area of the coil. The current passing through the coil can be adjusted to precisely control the intensity of the applied magnetic field. The magnetic signals on the surface of specimen No. 1 were only excited by the geomagnetic field. For specimen No. 2, the current passing through the Helmholtz coil was set as the fundamental current $I_B = 0.05$ A, which can produce a magnetic field whose intensity was $H_B = 40$ A/m (i.e., the intensity of the geomagnetic field). The current passing through the Helmholtz coil was adjusted to $1.5I_B$, $2I_B$ and $2.5I_B$ for specimens No. 3, 4 and 5, and they produced a series of magnetic fields with intensities of $1.5H_B$, $2H_B$ and $2.5H_B$, respectively. The magnetic signals $H_p(y)$ along the scanning lines from A to B were also measured by an EMS 2003 metal magnetic memory device with the same measurement parameters as shown in Sect. 8.2. The variations in the magnetic signals measured on the 5 testing lines have identical features; therefore, only the results from line 3 are presented and analyzed.

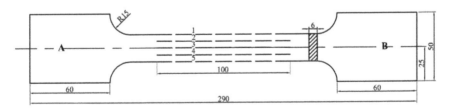

Fig. 8.7 Specimen shape and detection lines (in mm)

8.3.1 Magnetic Signals Excited by the Geomagnetic Field

The $H_p(y)$ curves excited by the geomagnetic field under different tensile stresses are shown in Fig. 8.8. The direction of the magnetic signals after loading is reversed, and all the measured curves are approximately linear, crossing at a distance of 50 mm, where the value of $H_p(y)$ is approximately zero. Thus, the left signals are positive, while the right signals are negative. Furthermore, the specimens show a magnetic ordering state to some extent after loading, in that the $H_p(y)$ curves rotate around their center in a clockwise direction with increasing stress in the elastic deformation stage. The $H_p(y)$ signals increase with increasing stress and reach their maximum under yield stress. However, the variation in the $H_p(y)$ signals in the elastic deformation stage is different from that in the plastic deformation stage, and it decreases slightly with further increasing stress until the specimens fail, as shown in Fig. 8.8b.

Given that linear magnetic signals are easily observed, it is feasible to extract the slope coefficient, K_S, of the $H_p(y)$ curves at different stresses for further investigation. The relationship between K_S and the applied stress is given in Fig. 8.9. The figure shows that the slope coefficient K_S is approximately linear to the applied stress σ in the elastic deformation stage, which is described by

$$K_S = a\sigma + b \tag{8.4}$$

where a and b are both constant values related to the applied stress σ, the material and the shape of the specimen. In the experiment, a is 1.00 and b is 0.09. The slope coefficient K_S and applied stress σ are well fitted by Eq. (8.4) in Fig. 8.9. The slope coefficient K_S reaches a maximum of 4.36 when tensile stress is applied under the yield stress and then decreases in the plastic deformation stage.

It is known that the magnetic domains turn along the axial direction of the applied load based on the piezomagnetic effect, leading to a magnetic behavior change in

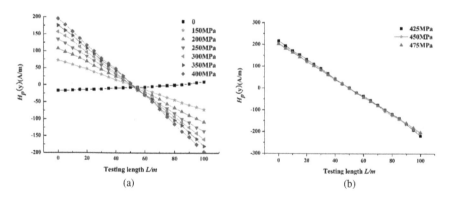

Fig. 8.8 $H_p(y)$ components of the residual magnetic field from measured lines excited by the geomagnetic field in the **a** elastic deformation stage and **b** plastic deformation stage

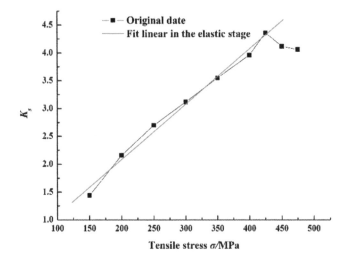

Fig. 8.9 Relationship between the slope coefficient K_S and the applied static tensile stress σ

ferromagnetic materials. Consequently, the specimen is magnetized along the direction of axial stress and is similar to a weak magnet with opposite polarities of the magnetic charges at its two ends. Therefore, the $H_p(y)$ signals on the surface of the specimen present good linearity after axial stress is applied. More magnetic domains are reoriented with increasing applied stress; thus, the magnetization of the specimen becomes increasingly stronger, represented by the linear increase in the amplitude of $H_p(y)$ and its slope coefficient K_S in the elastic deformation stage. The magnetic behavior of ferromagnetic materials in the plastic deformation stage is very complex. The dislocation density increases slowly, and dislocations are heterogeneously distributed with increased loads in the plastic deformation stage. Dislocations can pin the magnetic domain wall, creating some obstacles to the movement of magnetic domains, which is one reason for the reduction in the magnetic signals during the plastic deformation stage. In addition, residual compressive stress after unloading has a great effect on magnetic signals. The specimen exhibits tensile elongation after the applied stress is removed in the plastic deformation stage, which leads to residual compressive stress occurring in the specimen. During the plastic deformation stage, the orientation of the magnetization moment is changed to a direction perpendicular to the compressive stress; thus, the magnetic signals induced by stress decrease. The phenomenon that the slope coefficient K_S slowly decreases with increasing stress in the plastic deformation stage could be clarified by means of dislocations and residual compressive stress on magnetic domains. The tensile stress acting on the ferromagnetic materials can be calculated by the curve of K_S; therefore, the stress distribution in the ferromagnetic materials can be characterized by the slope coefficient K_S of the $H_p(y)$ curves because of its substantial dependence on stress.

8.3.2   Magnetic Signals Excited by the Applied Magnetic Field

The variation in the magnetic signals excited by different applied magnetic fields in the elastic deformation stage is shown in Fig. 8.10. These $H_p(y)$ curves vary similarly to the magnetic signals excited by the geomagnetic field. The $H_p(y)$ curves present good linearity along the testing lines and cross zero values at the position of 50 mm. The applied magnetic field affects the magnitude of the magnetic signals instead of the signal profile, and the $H_p(y)$ signals increase with increasing applied magnetic field or tensile stress. The $H_p(y)$ signals increased with increasing applied stress under the same applied magnetic field, and they increased with increasing applied magnetic field when the same stress is applied.

The variation in the magnetic signals excited by different applied magnetic fields in the plastic deformation stage is shown in Fig. 8.11. The $H_p(y)$ signals increase with increasing applied magnetic field when the same stress is applied (450 and 475 MPa) during the plastic deformation stage.

Figure 8.12 shows the variation in the slope coefficient K_S of the $H_p(y)$ curves under different applied magnetic fields with stress in the elastic deformation stage.

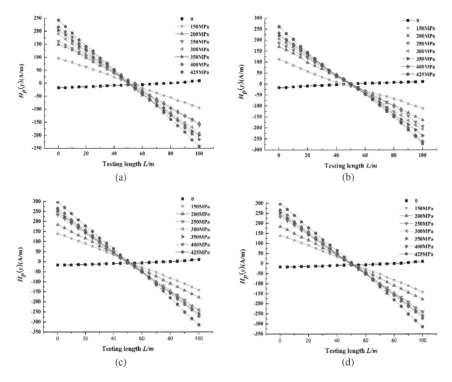

Fig. 8.10 $H_p(y)$ components of the magnetic signals from the testing line excited by the applied field with different magnetic field intensities: **a** H_B; **b** 1.5 H_B; **c** 2 H_B; and **d** 2.5 H_B

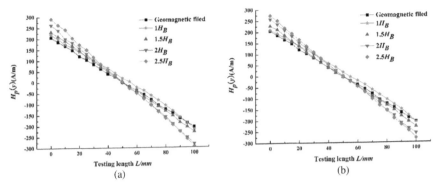

Fig. 8.11 $H_p(y)$ components of the magnetic signals from the testing lines excited by the applied field with different magnetic field intensities: **a** 450 MPa and **b** 475 MPa

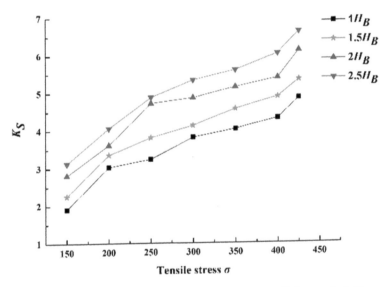

Fig. 8.12 Relationship between K_S and stress σ under different applied magnetic fields

The variation in K_S under different applied magnetic fields is similar to its variation under the geomagnetic field, and it increases with the increase in either the applied magnetic field or tensile stress. The change rate of the slope coefficient K_S, K_H, with the applied magnetic field H in the elastic deformation stage is given in Fig. 8.13. This shows that K_H has an approximately linear increase with the applied magnetic field H, which is described by

$$K_H = cH + d \tag{8.5}$$

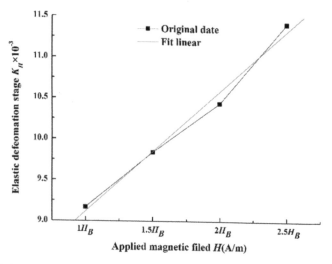

Fig. 8.13 Relationship between K_H and the applied magnetic field H

where c and d are also constant values related to the applied magnetic field, the materials and the shape of the specimen. In the experiment, c is 1.46 and d is 7.65. To further describe the variation in the magnetic signals with the applied magnetic field, the maximum gradient of the $H_p(y)$ curves on the testing lines, K_{max}, is given by

$$K_{max} = max(|\Delta H_p(y)/\Delta L|) \qquad (8.6)$$

where $\Delta Hp(y)$ is the maximum differential value of the magnetic signals between two points and ΔL is the distance between the two points.

Figure 8.14 shows the variation in the maximum gradient K_{max} with the applied magnetic field H at different tensile stresses. K_{max} increases with increasing applied magnetic field at the same stress, and it increases with increased stress when the same magnetic field is applied. The maximum gradient K_{max} can be used to identify the stress concentration that increases with increasing applied magnetic field. Thus, the use of an applied magnetic field can eliminate stray signals and help to evaluate the stress concentration by highlighting the characteristic values of the magnetic signals.

Based on the experimental results, the measured magnetic field signals can be affected by the applied magnetic field and stress. And the magnetic field signals are also related with the magnetic induction and the magnetic permeability of the specimen. The magnetic permeability μ, which is a physical quantity that can be used to characterize the magnetism of a magnetic medium, is a variable that is sensitive to external environments. When the ferromagnet is affected by a weak magnetic field and tensile stress, it can be observed that [2]

$$\mu = \mu_T(1 + bH/\mu_T)\left[a_0 + a_1|\sigma|^m exp(n|\sigma|)\right] \qquad (8.7)$$

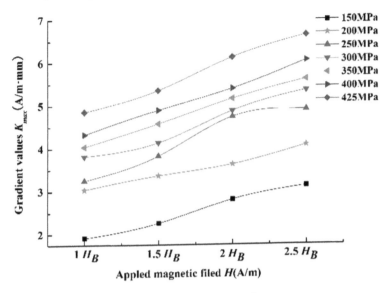

Fig. 8.14 Relationship between K_{max} and H at different tensile stresses

where T is the temperature, μ_T is the initial magnetic permeability relating to T, b is a constant relating to the material properties, and a_0, a_1, m, and n are coefficients depending on the direction and value of the applied stress. Equation (8.7) shows that the relation between the magnetic permeability and stress is nonlinear, including the power function and exponential function. Accordingly, the magnetic permeability increases rapidly, and the ferromagnet becomes magnetized when the stress increases, which agrees with the change in the slope coefficient K_S induced by tensile stress in the elastic deformation stage in the experiment.

The magnetic permeability curve of the carbon steel was measured by MATS-2010H (a permanent magnetic measuring device produced by LINKJOIN Corporation in China), as shown in Fig. 8.15. The permeability value increases sharply when the applied magnetic field increases from zero to approximately the threshold value (i.e., 280 A/m), and then it decreases to a constant value with an increased applied magnetic field. From the viewpoint of magnetic charge, a large amount of positive and negative magnetic charge gathers at the two interfaces of the specimen after loading and forms the N and S poles, respectively. Consequently, the specimen becomes an internal magnetic source, which tends to scatter the field outward. The normal component of the magnetic signals is positively correlated with the magnetic charge density ρ, which can be described as:

$$\rho = \mu \times M = \mu \times \left(\frac{\mu}{\mu_0} - 1\right)H \qquad (8.8)$$

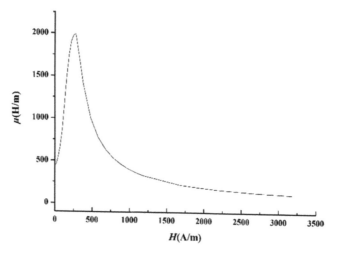

Fig. 8.15 Magnetic permeability curves of the carbon steel

In the experiment, the applied magnetic field intensity H was less than the threshold value, which led to an increase in μ with increased applied magnetic field intensity. It can be inferred that ρ increased with increasing applied magnetic field; thus, the result in the experiment shows that the magnetic signals on the surface of the specimen increase with increasing applied magnetic field H, which is well fitted by Eq. (8.8).

8.4   Conclusions

The applied magnetic field affects the magnitude of the MMM signal instead of changing the signal curve profile, and the MMM signal and its gradient increase with increasing applied magnetic field intensities, while the gradient is uncorrelated with the stress cycles. The $H_p(y)$ curve becomes more coincident with increasing applied excitation magnetic field. The maximum gradient K_{max} can be used to identify the stress concentration, which rises with the increase in either the applied magnetic field intensities or stress cycles. Furthermore, the change rates of the slope coefficients, K_S and K_H, also increase linearly with increasing applied magnetic field intensities under static tensile stress. This indicates that the applied magnetic field can effectively eliminate the interference induced by a stray magnetic field and help to strengthen the residual magnetic signals by highlighting its characteristic values to facilitate the evaluation of stress concentrations in ferromagnetic materials.

References

1. L.H. Dong, B.S. Xu, S.Y. Dong et al., Stress dependence of the spontaneous stray field signals of ferromagnetic steel. NDT&E Int. **42**(4), 323–327 (2009)
2. C.L. Shi, S.Y. Dong, B.S. Xu et al., Metal magnetic memory effect caused by static tension load in a case-hardened steel. J. Magn. Magn. Mater. **322**(4), 413–416 (2010)
3. K. Yao, Z.D. Wang, B. Deng et al., Experimental research on metal magnetic memory method. Exp. Mech. **52**(3), 305–314 (2012)
4. L.Q. Zhong, L.M. Li, X. Chen, Magnetic signals of stress concentration detected in different magnetic environment. Nondestruct. Test. Eval. **25**(2), 161–168 (2010)
5. S.K. Ren, Y.L. Yang, T.H. Lin, Influences of environment magnetic field on stress-magnetism effect for 20 steel ferromagnetic items. Phys. Test. Chem. Anal. (Part A: Phys. Test.) **2008**(8): 401–404 (2008)
6. D. Wang, B.S. Xu, S.Y. Dong, Discussion on background magnetic field control in metal magnetic memory testing. Nondestruct. Test. **29**(2), 71–73 (2007)
7. W.Y. Gong, Z.M. Wu, H. Lin et al., Longitudinally driven giant magneto-impedance effect enhancement by magneto-mechanical resonance. J. Magn. Magn. Mater. **320**(8), 1553–1556 (2008)
8. J.C. Leng, M.Q. Xu, M.X. Xu et al., Magnetic field variation induced by cyclic bending stress. NDT&E Int. **42**(5), 410–414 (2009)
9. J.W. Li, M.Q. Xu, J.C. Leng et al., Metal magnetic memory effect caused by circle tensile-compressive stress. Insight-Non-Destruct. Test. Cond. Monit. **53**(3), 142–145 (2011)
10. E.N. Yang, L.M. Li, X. Chen, Magnetic field aberration induced by cycle stress. J. Magn. Magn. Mater. **312**(1), 72–77 (2007)

Part III
Evaluation of the Repair Quality
of Remanufacturing Samples
by the MMM Technique

Chapter 9
Characterization of Heat Residual Stress During Repair

9.1 Introduction

After the damaged remanufacturing cores are evaluated, the cladding coating prepared by advanced surface engineering techniques will be deposited on the damage zones of the cores if they have remanufacturing value. As a typical remanufacturing repair technique, plasma transferred arc welding (PTAW) is commonly employed to prepare a cladding coating on a metallic substrate, which exhibits special advantages including a low dilution rate, intensive energy density, and high deposition speed. The metal powders are melted immediately when they pass through the plasma beam between the positive and negative electrodes. PTAW has been widely applied in equipment remanufacturing in the petrochemical industry and for engineering machinery and mining machinery [1] since the deposition of metal powders over the damaged region can greatly improve the wear resistance and corrosion resistance of components. However, due to the extremely high temperature of the plasma-transferred arc column, the metal powders undergo rapid melting and cooling processes. Therefore, the substrate materials will inevitably induce different degrees of structural transformation and uneven elastic-plastic deformation after remanufacturing. The variations in the thermal shrinkage and temperature gradient are very complex. During PTAW, the residual stress present in the welding area and heat-affected zone (HAZ) is complicated and inevitably attributed to the combined actions of plastic deformation, temperature gradients, and metallurgical changes [2]. The residual stress can greatly reduce the fatigue strength, induce crack initiation and even lead to serious accidents, which affect the service performance and life of remanufactured components [3]. Therefore, rapidly and accurately evaluating the residual stress has been the key to guaranteeing the quality of remanufacturing components repaired by PTAW. The evaluation results can directly affect the safety performance of remanufactured products in the service process.

Previous studies indicate that the spontaneous magnetic signals have the potential to evaluate the early damage degree and forecast the residual service life of ferromagnetic structural components [4–7]. Considerable research on welding residual

© Science Press 2021

H. Huang et al., *Metal Magnetic Memory Technique and Its Applications in Remanufacturing*, https://doi.org/10.1007/978-981-16-1590-0_9

stress also has been carried out via spontaneous magnetic signals. Qi et al. [8] investigated the stress distribution of welded seams on CCSB ship plates by using metal magnetic memory (MMM) technique, and the results showed that the region with high residual stress can be rapidly located. Liang et al. [9] determined that the residual stress that was qualitatively dependent on the gradient of the magnetic signal $H_p(y)$ perpendicular to the welded seam. Dubov et al. [10, 11] assessed the stress state of welded joints by using MMM because they recognized that the stress concentration zones are sources of which pipeline damage emerges and develops [12]. In addition, a series of welding studies conducted by Roskosz showed that the nonuniform distribution of welded residual stresses resulted in variations in the magnetic signals [13], which could be used to assess the level of residual stress in welded seams [14]. In addition to welded seams, Liu et al. [15] also evaluated the residual stress in laser cladding coatings. The slope of the magnetic signal $H_p(y)$ increased with increasing stress but decreased when the stress approached the yield strength. Similarly, Huang et al. [16] detected the stress of a PTAW cladding coating and found that the gradient of magnetic signal K increased with an increase in stress at the elastic stage, while it decreased when the deformation entered the plastic stage. In addition, the hysteresis curves of the PTAW cladding coating were also analyzed to explain the change mechanism of magnetic signals with stress. Previous studies of MMM primarily focused on the stress evaluation of welded seams and cladding coatings but rarely addressed HAZs. In fact, residual stress can also occur at the HAZ where the temperature exceeds the critical values for phase transformations [2], and the stress concentration zones are usually the locations of defects or material property degradation. Accordingly, residual stress evaluation for the HAZ is an urgent problem.

In this chapter, the variations in the spontaneous magnetic signal tangential components $H_p(x)$ and normal components $H_p(y)$ are studied to reflect the residual stress in the PTAW cladding coating and HAZ. The distribution and level of residual stress are further characterized by the newly proposed spatial magnetic signal characteristic value H_p and its average value H_{pa}, respectively. The chapter is organized as follows. In Sect. 9.2, the magnetic signal measurement is carried out for the PTAW cladding coating, and the experimental data are processed. The MMM signal measurement results are shown and analyzed in Sect. 9.3, including magnetic signals parallel and perpendicular to the cladding coating and three-dimensional spatial magnetic signals. To validate the reliability of the applied MMM method, the residual stress in the welded specimen is analyzed by X-ray diffraction (XRD). To determine the generation mechanisms of high residual stress and a large magnetic field in the HAZ, the microstructure and microhardness are also investigated in Sect. 9.4. This research can provide guidance for the quality control of ferromagnetic remanufacturing components repaired by PTAW.

Table 9.1 Chemical composition of Ni60 feed powder (wt%)

Material	C	Cr	Si	B	Fe	Ni
Ni60	0.6–1.0	14–17	3.0–4.5	2.5–4.5	≤15	Bal.

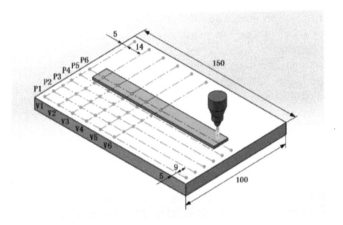

Fig. 9.1 Size and location of the PTAW cladding coating and scanning lines

9.2 Preparation of Cladding Coating and Measurement of MMM Signals

9.2.1 Specimen Preparation

Ferromagnetic Q345 structural steel with dimensions of 150 × 100 × 10 mm was chosen as the substrate of PTAW due to its good mechanical and welding properties. Nickel-based self-fluxing alloy (Ni60) was used as the feed powder, whose chemical composition is shown in Table 9.1. The size and location of the PTAW cladding coating is given in Fig. 9.1. Since the cladding coating is centrosymmetric about the midpoint of the specimen, only one side of the residual stress distribution on the specimen is investigated. Six scanning lines, marked by P1–P6, are arranged along the cladding coating, while the other six scanning lines, marked by V1–V6, are perpendicular to the cladding coating. All these scanning lines have thirty-six intersection points.

9.2.2 Measurement Method

To eliminate the effect of mechanical processing and transportation on the initial magnetic signals, the substrate was polished and demagnetized before PTAW. Then, the substrate was placed on a nonmagnetic material platform along the South-North direction. During the detection process, the initial magnetic signals were measured on each scanning line by a TSC-2M-8 device. The measurement system was made by Energodiagnostika Co. Ltd. The probe with a 1 A/m sensitivity based on the Hall sensor was gripped on a nonferromagnetic scanning device and was placed vertical to the surface of the specimen with a lift-off value of 5 mm and horizontal movement speed of 8 mm/s. The probe can simultaneously collect the tangential components and normal components of the magnetic signal with a scanning increment of 1 mm along the scanning direction. The initial stress states in the 36 intersection points were also detected by XRD before PTAW.

The cladding coating was deposited by PTAW on the substrate after the aforementioned detection procedure. The main processing parameters of PTAW are given in Table 9.2. The dimensions of the cladding coating are 120 × 14 × 4 mm without defects of porosity and cracking. By repeating the aforementioned detection method and procedure, the residual magnetic signals and residual stress states were also measured after the specimen was sufficiently cooled. Then, the specimen was sectioned along the V6 scanning line, and the cross-section is shown in Fig. 9.2. The microstructures of the cladding coating, HAZ, and substrate were observed by metallographic microscopy. Before examination, the cross-section was polished and etched in 4% nital. The microhardness was also measured along the testing points by a low load Vickers hardness tester with the hardness gradient test method, as shown in Fig. 9.2. In the experiment, the specifications of the MMM measurement system, XRD, and microhardness are listed in Table 9.3, Table 9.4, and Table 9.5, respectively.

Table 9.2 Processing parameters of PTAW

Parameters	Unit	Value
Plasma arc current	A	100
Plasma arc voltage	V	26
Welding speed	mm/min	65
Powder feeding rate	g/min	18

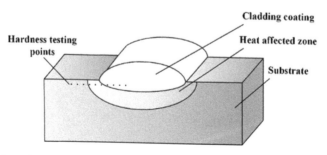

Fig. 9.2 Schematic diagram of the cross-section

Table 9.3 Specification of the MMM measurement system

Parameters	Unit	Value
Sensitivity	A/m	1
Lift-off	mm	5
Movement speed	mm/s	8
Scanning increment	mm	1

Table 9.4 Specification of the XRD measurement system

Parameters	Unit	Value
Accuracy	MPa	±8
Penetration depth	μm	30
Measurement point size	mm	0.5
Testing time	min	3

Table 9.5 Specification of the microhardness measurement system

Parameters	Unit	Value
Sensitivity	HV	1
Applied load	N	9.8
Loading time	s	15
Dwell time	s	10

9.2.3   Data Preprocessing

As a kind of spontaneous weak magnetic signal, the measurement accuracy of MMM is easily affected by the external environment, and data fluctuation may occur during testing. To reduce the testing error, the magnetic signals on each scanning line were repeatedly collected at three different times, and their average value was considered to be the final result. The scattering of the data was characterized by standard error bars. All the testing errors are very small, which indicates that the magnetic signals are stable and available for further analysis.

In this chapter, the variations in the magnetic signal tangential components $H_p(x)$ and normal components $H_p(y)$ are given as:

$$H_p(x) = H_{p2}(x) - H_{p1}(x) \tag{9.1}$$

$$H_p(y) = H_{p2}(y) - H_{p1}(y) \tag{9.2}$$

where $H_{p1}(x)$ and $H_{p2}(x)$ denote the collected magnetic signal tangential components before and after PTAW, respectively, and $H_{p1}(y)$ and $H_{p2}(y)$ denote the collected magnetic signal normal components before and after PTAW, respectively. The main interest in this chapter is on the change laws of magnetic signals under the effect of

PTAW, rather than the intensity of magnetic signals collected before or after PTAW. Therefore, we pay more attention to researching the differential value of the magnetic signals induced by PTAW in the following sections.

9.3   Distribution of MMM Signals Near the Heat Affected Zone

9.3.1   Magnetic Signals Parallel to the Cladding Coating

Based on the aforementioned data processing, the magnetic signals on lines P1–P6 parallel to the cladding coating are given in Fig. 9.3. The magnetic signal tangential components $H_p(x)$ formed U-shaped curves, as shown in Fig. 9.3a. From lines P1 to P5, the valley value of $H_p(x)$ increases with decreasing distance from the cladding coating. Similarly, it can be seen from Fig. 9.3b that the normal component curves rotate clockwise, which means that the intensity of $H_p(y)$ and the absolute value of its gradient also increases with decreasing distance from the cladding coating. However, both the valley value of $H_p(x)$ and the absolute value of the gradient of $H_p(y)$ on line P6 start to decrease compared with other scanning lines. This indicates that the magnetic signals on the cladding coating are smaller than those on the HAZ.

In addition, both the valley value of $H_p(x)$ and the zero value of $H_p(y)$ are located in the range of approximately 80–100 mm. Based on the evaluation criteria from Ref. [17], we can deduce that high residual stress is also mainly concentrated in this range. It is interesting that the high stress concentration zone deviates from the middle position of the scanning lines. One of the main mechanisms that generates residual stress is the temperature gradient [2]. Understanding the temperature distribution in the PTAW process can be of great importance for studying variations in magnetic

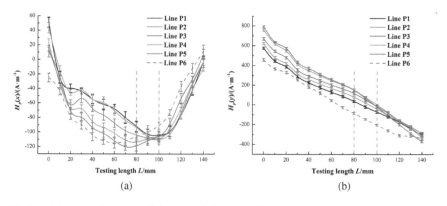

Fig. 9.3 Magnetic signals parallel to the cladding coating: **a** tangential components $H_p(x)$ and **b** normal components $H_p(y)$

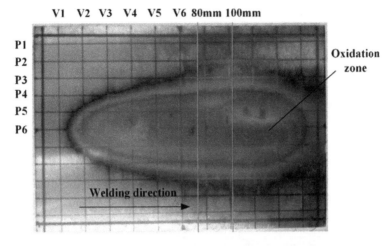

Fig. 9.4 Spindle-shaped welding oxidation zone on the back of the substrate

signals and stress. At the beginning of PTAW, very little heat is carried to the substrate because of the fluctuations in powder delivery and the instability of the plasma beam. As the PTAW continues, the heat increases continuously, and a sharp temperature gradient is present in the latter portion of the specimen, where a higher residual stress and wider welding oxidation zone appear. Therefore, the maximum width zone of welding oxidation is also located in the range of approximately 80–100 mm on the back of the substrate, as shown in Fig. 9.4. This indicates that the oxidation zone induced by high temperatures can reflect the location of high residual stress and magnetic signal features such as the valley value of $H_p(x)$ or the zero value of $H_p(y)$.

9.3.2   Magnetic Signals Perpendicular to the Cladding Coating

Analogously, the variations in the magnetic signals perpendicular to the cladding coating are given in Fig. 9.5. The magnetic signal tangential components $H_p(x)$ on lines V2–V6 show peak troughs in Fig. 9.5a. To better describe the variations in the magnetic signals under the effect of PTAW, the actual distribution of the deposition track on the specimen is displayed in Fig. 9.6. The cladding coating is located at the middle position of the specimen, and its width is approximately 14 mm. Based on this, the red dashed lines, which can represent the actual position and width of the cladding coating, are marked in Fig. 9.5. The overall trends of the distance between the peak and trough and its middle position almost coincide with the width and position of the cladding coating. In addition, it can be seen from Fig. 9.5b that the slopes of the normal component of the magnetic signals $H_p(y)$ increase gradually when they enter the cladding coating zone and reach a maximum at the HAZ. Meanwhile, the overall

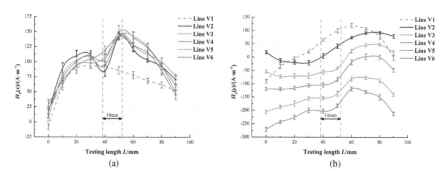

Fig. 9.5 Magnetic signals perpendicular to the cladding coating: **a** tangential components $H_p(x)$; and **b** normal components $H_p(y)$

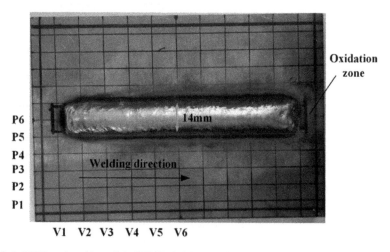

Fig. 9.6 Width and position of the PTAW cladding coating

$H_p(y)$ curves move down, which means that the overall intensity of the magnetic signals increases along the deposition track movement direction. The residual stress on line V2 is the lowest because of the low energy input at the beginning of PTAW, as previously mentioned. In contrast, the residual stress on line V6 is the highest, which is attributed to more energy input with the progress of PTAW. Furthermore, the variation in the magnetic signals on line V1 without passing through the cladding coating is obviously different from other scanning lines. It is impossible to analyze the information about the width and position of the cladding coating based only on the detection result on line V1.

It is worth noting that the magnetic memory signals are susceptible to numerous interference factors, which can result in substantial difficulties for signal feature extraction. The variations in the real measured magnetic signals, as shown in Fig. 9.5, are not in accordance with the ideal results in Ref. [7]. The tangential components

$H_p(x)$ may only reach a maximum value in the local high residual stress region rather than the global area. The zero-crossing position of the normal components $H_p(y)$ may not have a certain correlation with the stress concentration. This indicates that it is difficult to reflect the stress distribution with only two conventional characteristic values $H_p(x)$ or $H_p(y)$. A new characteristic value needs to be introduced to further characterize the residual stress of PTAW in the rest of this chapter.

9.3.3 Three-Dimensional Spatial Magnetic Signals

To better reflect the distribution of PTAW residual stress, the three-dimensional spatial magnetic signals are given as follows:

$$H_p = \pm\sqrt{\left[H_{pp}(x)\right]^2 + \left[H_{pv}(x)\right]^2 + \left[(H_{pp}(y) + H_{pv}(y))/2\right]^2} \qquad (9.3)$$

where $H_{pp}(x)$ and $H_{pp}(y)$ are the magnetic signal tangential components and normal components parallel to the cladding coating, respectively. $H_{pv}(x)$ and $H_{pv}(y)$ are the magnetic signal tangential components and normal components perpendicular to the cladding coating, respectively. The projection of the spatial magnetic signal characteristic value H_p in the three-dimensional Cartesian coordinate system is shown in Fig. 9.7. The average values of $H_{pp}(y)$ and $H_{pv}(y)$ are considered components of the Z axis. The characteristic value H_p is positive when it points in the positive direction of the Z axis and negative when it points in the opposite direction. The characteristic values H_p at thirty-six intersection points are calculated, and their distribution nephogram is given in Fig. 9.8 based on the two-dimensional discrete interpolation method.

For the scanning lines parallel to the cladding coating, it can be observed from Fig. 9.8 that the gradient of the spatial magnetic signals H_p on line P5 reaches the maximum along the arrow marked at HAZ; however, the intensity of spatial magnetic

Fig. 9.7 The projection of H_p in a three-dimensional Cartesian coordinate system

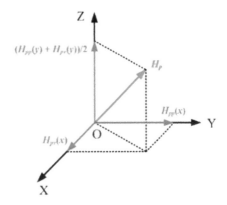

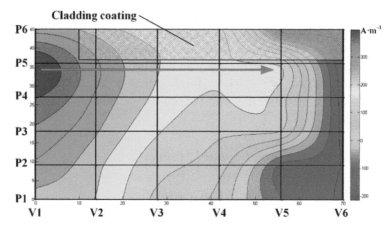

Fig. 9.8 Distribution nephogram of three-dimensional spatial magnetic signals H_p

signals H_p decreases. Based on Ref. [18], the specimen can be considered as a whole piece of a magnetic block with positive and negative magnetic poles distributed on the two sides of the specimen, respectively. A schematic diagram of the magnetic induction line distribution on the surface of the specimen in PTAW is shown in Fig. 9.9. A large number of magnetic induction lines leak out from the left edge of the specimen where a larger magnetic flux density is present. Therefore, the intensity of the spontaneous magnetic field decreases when the magnetic induction lines are far away from the left edge of the specimen. For the scanning lines perpendicular to the cladding coating, the intensity of the spatial magnetic signals H_p along lines V1–V6 also reaches a maximum at the HAZ. As is well known, the residual stress is mainly concentrated near the interface of the coating. Therefore, the dramatic increase in

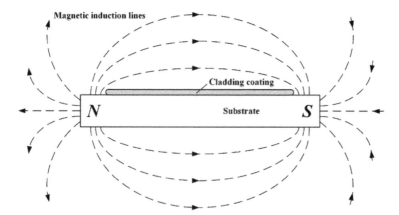

Fig. 9.9 Schematic diagram of the magnetic induction line distribution on the surface of the specimen

the spatial magnetic signals H_p near the coating may be induced by the high residual stress in the HAZ.

9.3.4 Verification Based on the XRD Method

To verify the reliability of MMM, the spatial magnetic signals H_p are compared with the residual stress measured by the XRD method. The detection results parallel to the cladding coating are shown in Fig. 9.10. The XRD residual stress shows an increasing trend on lines P1–P6 because more heat and a higher temperature gradient are concentrated at the latter portion of the testing line as the plasma beam moves forward during the PTAW process. However, the spatial magnetic signal value H_p on lines P1–P6 shows a decreasing trend. The phenomenon has been explained in Sect. 9.3.3 by assuming the specimen is a magnetic block, which leads to a large amount of magnetic flux density concentrated at the front portion of the testing line. The detection results perpendicular to the cladding coating are also given in Fig. 9.11. The maximum values of the residual stress and spatial magnetic signals H_p are both located at the P5 position (i.e., the HAZ next to the cladding coating), except for V6. The individual detection result does not follow the change law, which may be due to signal interference from external factors. The deviation of individual detection results may not affect the overall variation trends of the residual stress or magnetic signals.

According to the magnetomechanical model, the variation in stress can lead to a change in spontaneous magnetization. Some studies have been carried out to successfully establish the quantitative relationship between stress and magnetic signals. Furthermore, it seems that the spatial magnetic signal is relevant to the residual stress in PTAW based on the aforementioned analysis. The average values of the residual stress σ_a and spatial magnetic signals H_{pa} obtained by XRD on each scanning line are defined as follows:

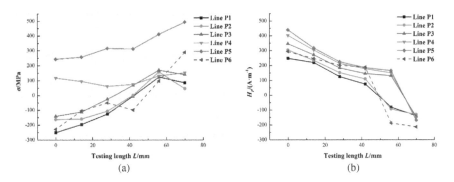

Fig. 9.10 Detection results parallel to the cladding coating: **a** XRD residual stress and **b** spatial magnetic signals

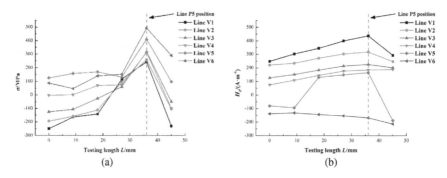

Fig. 9.11 Detection results perpendicular to the cladding coating: **a** XRD residual stress; and **b** spatial magnetic signals

$$\sigma_a = \frac{1}{I} \sum_{i=1}^{I} \sigma_i \tag{9.4}$$

$$H_{pa} = \frac{1}{I} \sum_{i=1}^{I} H_{pi} \tag{9.5}$$

where σ_i is the residual stress of the ith data point measured by XRD, H_{pi} is the spatial magnetic signal of the ith data point calculated based on Eq. (9.3), and I = 6 is the number of data points. The variations in the average values of the XRD residual stress σ_a and spatial magnetic signals H_{pa} on each scanning line are plotted in Fig. 9.12. It is believed that H_{pa} parallel to the cladding coating is positively related to σ_a except for line P6, while H_{pa} perpendicular to the cladding coating is negatively related to σ_a. One key issue in MMM applied to PTAW is establishing the quantitative relationship between the residual stress and magnetic signal characteristic value. As shown in Fig. 9.13a, an approximate exponential relationship is found between

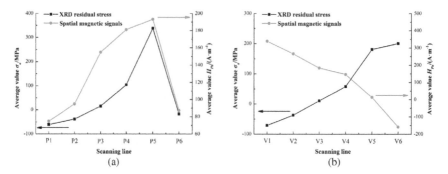

Fig. 9.12 Variations in σ_a and H_{pa}: **a** parallel to the cladding coating and **b** perpendicular to the cladding coating

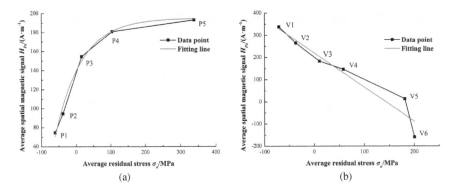

Fig. 9.13 Relationship between σ_a and H_{pa}: **a** parallel to the cladding coating and **b** perpendicular to the cladding coating

σ_a and H_{pa} parallel to the cladding coating, which can be fitted by the following exponential fitting equation:

$$H_{pa} = c_1 + a_1 \times \exp(b_1\sigma_a) \tag{9.6}$$

where a_1, b_1, and c_1 are constant values related to the processing parameters of PTAW. In this experiment, a_1 is -55.368, b_1 is -0.013, c_1 is 194.776, and R^2 (adjusted deviate square) is 0.980. The growth rate of the magnetic signals decreases with an increase in the residual stress when approaching the P5 position. This is responsible for the very large quantitative differences between the residual stress and magnetic signals, especially for line P5 in Fig. 9.10. For testing perpendicular to the cladding coating, a linear relationship between σ_a and H_{pa} is plotted in Fig. 9.13b, and the fitting equation is listed as follows:

$$H_{pa} = a_2 + b_2 \times \sigma_a \tag{9.7}$$

where a_2 and b_2 are constant values that are also affected by the processing parameters of PTAW. In this equation, a_2 is 218.587, b_2 is -1.542, and R^2 is 0.915. Good agreement is achieved between the spatial magnetic signals and residual stress measured by XRD, which shows that the characteristic value H_p and its average value H_{pa} have the capacity to reflect the distribution and level of residual stress in PTAW, respectively.

9.4   Generation Mechanism of MMM Signals in the Heat Affected Zone

9.4.1   Microstructure and Phase Transformation

The microstructures of the cladding coating, HAZ, and substrate are observed to explore the effect of PTAW on the distribution of spontaneous magnetic signals and residual stress. When a high energy plasma beam irradiates the molten pool in PTAW, the temperature gradient between the bottom of the molten pool and the surface provides a driving force for the growth of the grain. The crystal development of austenite, considered as the initial phase, is significantly promoted because the wide subcooled region in front of liquid-solid interface brings the sufficient time for the grain growth. However, when the plasma beam moves out of the molten pool and the temperature is lower than Ar3, the primary austenite phase partially transforms into ferrite or pearlite due to the high solidification rates. Finally, it can be seen from Fig. 9.14a that the obvious dendrites with residual austenite, acicular ferrite, and Widmanstatten ferrite are present in the cladding coating at room temperature. Besides, a large amount of carbide and boride with high hardness are also precipitated, which can help to improve the abrasive resistance of coating. As a kind of paramagnetic material, austenite can hardly be magnetized spontaneously. Therefore, the magnetic signal intensity of cladding coating contained partial austenitic structure is weaker than that of HAZ. This is responsible for not only the reduction

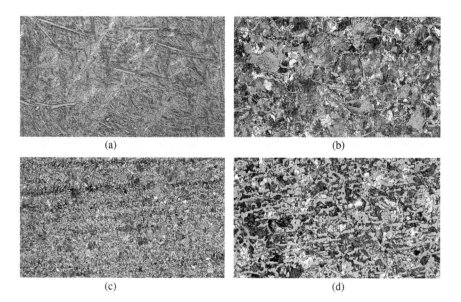

(a)   (b)

(c)   (d)

Fig. 9.14 Microstructure of the PTAW specimen: **a** cladding coating; **b** overheated zone; **c** recrystallization zone; and **d** substrate, magnification: 200×

of the valley value of $H_p(x)$ and the gradient absolute value of $H_p(y)$ on line P6 as shown in Fig. 9.3, but also the abrupt reduction of spatial magnetic signal H_p at line P6 position as shown in Fig. 9.11.

The HAZ can be classified as overheated zone and recrystallization zone. Their microstructure can be seen in Fig. 9.14b and c, respectively. We can find that the grain coarsening occurred in Fig. 9.14b where a large amount of pearlite is present in this area due to element diffusion of carbon and alloy. Besides, a certain amount of bainite appears attributed to high temperature and cooling rate. In Fig. 9.14c far away from heat source, the finer grains can be observed at the recrystallization zone composed of pearlite and ferrite. During the cooling process, the non-uniform shrinkage effect of weld seam occurs due to surrounding temperature gradients. Besides, the volumetric change at microscale is induced by these phase transformations in HAZ. They are both responsible for the increment of strain and residual stress. Based on the magnetomechanical effect, the residual stress in HAZ can lead to the spontaneous magnetic signals. It explains the reason for the distortion of magnetic signal components and larger spatial magnetic signals at HAZ, as mentioned before.

The structure of the substrate far away from the HAZ contains ferrite and pearlite grains, as shown in Fig. 9.14d, which is identical to the microstructure in the initial material before PATW. Thus, the residual stress in the substrate remains at a lower level because no phase transformation occurs.

9.4.2 Microhardness Distribution

The variations in the magnetic signals and residual stress in PTAW are both related to metallographic structure changes based on the aforementioned analysis. As a substantial evaluation indicator for welding quality, the microhardness is also strongly dependent on the metallographic structure. To better understand the change mechanism of magnetic signals and residual stress, it is useful to research the distribution of microhardness near the fusion line, and the testing results are shown in Fig. 9.15. The zero value position at the abscissa axis corresponds to the fusion line. The negative value at the abscissa axis represents the cladding coating zone, while the positive value represents the HAZ and substrate zones. The hardness of the cladding coating is 600–700 HV, which shows great abrasive resistance due to the strengthening phase of carbides or borides dispersed in the Ni-based alloy. When crossing the fusion line and entering the HAZ and substrate, the hardness decreases sharply to 250–300 HV because a large amount of pearlite and ferrite whose hardness is lower than the strengthening phase appears. The rapid decrease in hardness from the cladding coating into the HAZ and substrate indicates that various phases exist at the interface. The various phases with different material mechanical properties may lead to different amounts of volume shrinkage during the cooling process, which is responsible for the high strain or stress concentration near the bonding interface. Based on the magnetomechanical effect, the magnetic signal characteristic values H_p and H_{pa} also reach a maximum near the interface, as shown in Figs. 9.11b and 9.12a.

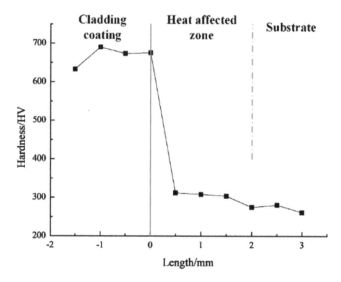

Fig. 9.15 Microhardness near fusion line

9.5 Conclusions

In this chapter, a novel magnetic nondestructive testing method based on spontaneous magnetic signals is used to characterize the residual stress in PTAW. According to both the magnetic signal tangential components $H_p(x)$ and normal components $H_p(y)$, the high residual stress zones can be located, and the amplitude of $H_p(x)$ and the slope of $H_p(y)$ can determine the location and size of the cladding coating. On this basis, a new characteristic value H_p and its average value H_{pa} have been proposed, which have a good capacity for reflecting the distribution and level of residual stress. The characteristic value H_{pa} increases exponentially with an increase in the average residual stress parallel to the cladding coating, whereas it decreases linearly perpendicular to the cladding coating. This MMM technique can potentially replace the existing destructive methods of residual stress determination and contribute to remanufacturing repair quality evaluation during PTAW. Compared with other nondestructive testing methods, MMM has a maximum measuring velocity of up to several meters per second without concerning surface preparation, so it can rapidly characterize the distribution and level of residual stress. Based on the horizontal movement speed of the probe and the length of the scanning line in this experiment, it only takes several minutes to collect all the magnetic signal data.

In addition, various phases exist at different zones of the specimen during the PTAW cooling process, and thus, a rapid decrease in hardness from the cladding coating into the HAZ occurs. Except for the temperature gradients, the various phases with different material mechanical properties may lead to different volume shrinkage during the cooling process, which is the possible reason for the high residual stress and large magnetic field at the HAZ. Although the level of residual stress in the HAZ may be different in various coating materials, such as Ni, Fe, and Co-based alloys, we can assume that these overall variation trends or distributions of the residual stress are the same because the phase transformation always occurs at the HAZ. Therefore, the MMM technique has wide applicability for characterizing the residual stress in PTAW during remanufacturing.

References

1. D.W. Deng, R. Chen, H.C. Zhang, Present status and development tendency of plasma transferred arc welding. J. Mech. Eng. **49**(7), 106–112 (2013)
2. H.I. Yelbay, I. Cam, C.H. Gür, Non-destructive determination of residual stress state in steel weldments by Magnetic Barkhausen Noise technique. NDT&E Int. **43**, 29–33 (2010)
3. L.H. Dong, Y. Wang, W.L. Guo, *Stress Measuring Techniques and Applications in Remanu-facturing Field* (National Defense Industry Press, Beijing, 2014)
4. W. Wang, S.C. Yi, S.Q. Su, Experimental investigation of stress and damage characterization of steel beam buckling using magnetic memory signals. Struct. Des. Tall Spec. Build. **25**(11), 505–518 (2016)
5. H.M. Li, Z.M. Chen, Quantitative analysis of the relationship between non-uniform stresses and residual magnetizations under geomagnetic fields. AIP Adv. **6**(7), 401–406 (2016)
6. M. Roskosz, M. Bieniek, Evaluation of residual stress in ferromagnetic steels based on residual magnetic field measurements. NDT&E Int. **45**, 55–62 (2012)
7. B.S. Xu, L.H. Dong, *Metal Magnetic Memory Testing Method in Remanufacturing Quality Control* (National Defense Industry Press, Beijing, 2015)
8. X. Qi, S. Di, H. Liu et al., Magnetic Barkhausen noise, metal magnetic memory testing and estimation of the ship plate welded structure stress. J. Nondestruct. Eval. **31**, 80–89 (2012)
9. B. Liang, J.M. Gong, H.T. Wang et al., Evaluation of residual stresses in butt-welded joints by residual magnetic field measurements. Appl. Mech. Mater. **217–219**, 2427–2434 (2012)
10. S. Kolokolnikov, A. Dubov, O. Steklov, Assessment of welded joints stress-strain state inhomogeneity before and after post weld heat treatment based on the metal magnetic memory method. Weld. World **60**, 665–672 (2016)
11. S. Kolokolnikov, A. Dubov, A. Marchenkov, Determination of mechanical properties of metal of welded joints by strength parameters in the stress concentration zones detected by the metal magnetic memory method. Weld. World **58**, 699–706 (2014)
12. A. Dubov, S. Kolokolnikov, Comprehensive diagnostics of parent metal and welded joints of steam pipeline bends. Weld. World **54**(9–10), 241–248 (2010)
13. M. Roskosz, Capabilities and limitations of using the residual magnetic field in NDT, in *19th World Conference on Non-Destructive Testing*, Munich (2016)
14. M. Roskosz, Metal magnetic memory testing of welded joints of ferritic and austenitic steels. NDT&E Int. **44**, 305–310 (2011)
15. B. Liu, S.J. Chen, S.Y. Dong et al., Stress measurement of laser cladded ferromagnetic coating with metal magnetic memory. Trans. China Weld. Instit. **36**(8), 23–26 (2015)
16. H.H. Huang, G. Han, C. Yang et al., Stress evaluation of plasma sprayed cladding layer based on metal magnetic memory testing technology. J. Mech. Eng. **52**(20), 16–22 (2016)
17. Z.D. Wang, Y. Gu, Y.S. Wang, A review of three magnetic NDT technologies. J. Magn. Magn. Mater. **324**(4), 382–388 (2012)
18. F.M. Gao, J.C. Fan, Research on the effect of remanence and the earth's magnetic field on tribo-magnetization phenomenon of ferromagnetic materials. Tribol. Int. **109**, 165–173 (2017)

Chapter 10
Detection of Damage in Remanufactured Coating

10.1 Introduction

For the remanufactured components whose heat residual stress meets the requirements, the service performance of the coatings also needs to be evaluated afterwards. The evaluation of remanufactured coatings is the key to guaranteeing subsequent safety and stability during the service process. We also take plasma transferred arc welding (PTAW) as an example because it has become a very important surface engineering technique and has been widely used in the remanufacturing field.

It is quite important to study the physical and mechanical properties after remanufacturing using the PTAW method. These properties include crack behavior induced by residual stress and temperature gradients and the impact of PTAW process parameters on the tensile properties and microstructure. Veinthal et al. analyzed the relation between the arc energy and the surface fatigue wear behavior of Fe-Cr-C powder [1]. Their study showed that different cooling times lead to differences in the microstructure and formation of residual stresses and surface cracks in surface fatigue wear testing. García-Vázquez et al. deposited nickel-based alloys on D2 steel and evaluated the metallurgical features on the weld beads/substrate interface by using scanning electron microscopy. They also examined the mechanical properties by wear tests [2]. Dikbas et al. used X-ray radiographic tests on Ti6Al4V alloy weld joints, and their results indicated that increasing the welding power increased the widths of deep penetration in all specimens [3].

Existing studies indicate that the defects and stress concentration in the cladding coating are key factors for evaluating the welding structure and predicting mechanical failure. Traditional nondestructive testing (NDT) methods, such as ultrasonic testing, radiographic inspection and eddy current testing, are widely used to test the defects of welded joints. These traditional NDT techniques can effectively detect macrocracks and defects, but they are incapable of determining early damage caused by stress concentrations [4]. Metal magnetic memory (MMM) signals have the potential to evaluate the heat residual stress of remanufacturing components, as shown in Chap. 9, and previous studies reported that $H_p(y)$ or $H_p(x)$ can be used to evaluate

© Science Press 2021

H. Huang et al., *Metal Magnetic Memory Technique and Its Applications in Remanufacturing*, https://doi.org/10.1007/978-981-16-1590-0_10

the degree of damage of materials in the elastic and plastic deformation stages [5, 6]. The MMM method for the detection of traditional engineered metal materials, such as 45 steel, Q345, and A283 Grade C mild steel, has been widely investigated [5, 6, 7, 8]. However, variances among magnetic signals and how they relate to the microstructure of PTAW remanufactured coatings under fatigue loads remain unknown. Similarly, the differences in the MMM signals between cladding coatings and a single ferromagnetic material (such as substrate) remain unknown. Furthermore, existing studies also fail to examine how MMM signals can be used to evaluate the surface quality of PTAW remanufactured coatings.

In this chapter, the MMM signals of a cladding coating and substrate under dynamic tensile tests were compared, and the $H_p(y)$ distributions on the surfaces of each specimen at different fatigue cycles were investigated. Vibrating sample magnetometer (VSM) tests were performed. The microstructures of the PTAW remanufactured coating were obtained to analyze the relationship between the MMM signals and microstructures. The remainder of this chapter is organized as follows. The fatigue test is carried out on the standard specimens of the PATW coating, and the corresponding MMM signals are extracted in Sect. 10.2. The relationship between the MMM signals and fatigue cycles shows that the slope of $H_p(y)$ and the average of $H_p(x)$ are potentially useful indicators for monitoring and evaluating the surface quality for PTAW remanufactured coatings under fatigue loads in Sect. 10.3. Finally, the conclusion is summarized in Sect. 10.4.

10.2   Cladding Coating and Its MMM Measurement

The substrate of the specimens was made of 45 steel. The chemical composition of this substrate is shown in Table 5.1. Ferrite-based self-fluxing alloy (Fe321 alloy), which has excellent mechanical performance, was used as the feed powder. The chemical composition of this alloy is shown in Table 10.1. The main processing parameters for PTAW used are given in Table 10.2.

After welding, the specimens were machined in accordance with the Chinese national standard GB/T 228-2002. The specimens had dimensions of 200 mm in length, 30 mm in width, 6 mm in thickness, and 3 mm in deposition thickness for the PTAW coating. Two small U-notches were carefully cut by electric discharge machining (EDM) at the center of each side of the specimen. Three scanning lines, labeled #1, #2 and #3, were drawn on the surface of the specimen, as shown in Fig. 10.1. Contrasting specimens made of 45 steel were also manufactured with the same geometry and dimensions. To eliminate the effect of welding thermal stress and

Table 10.1 Chemical composition of Fe321 powder (wt%)

Material	C	Cr	Si	B	Mo
Fe321	≤0.15%	12.5–14.5%	0.5–1.5%	1.3–1.8%	0.5–1.5%

Table 10.2 PTAW parameters

Parameters	Unit	Value
Welding current	A	150
Powder feeding voltage	V	20
Swing speed	mm/min	1800
Swing width	mm	35
Powder feeding amount	g/min	23

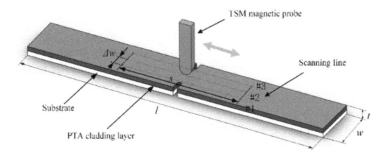

Fig. 10.1 The shape of the specimen and schematic diagram of the tests

mechanical processing on the initial magnetic field, the specimens were polished and demagnetized before loading. The specimens were annealed at 650 °C for 4 h in a vacuum electric furnace, followed by furnace cooling to room temperature.

Dynamic tension loads were applied to the specimens on an SDS 100 servo hydraulic testing machine. The dynamic load error was within ±1.0%. Tension–tension fatigue tests were performed for a constant amplitude (sinusoidal waveform) at a maximum load of 35 kN (stress at 328 MPa), minimum load of 3.5 kN (stress at 32.8 MPa) and load frequency of 10 Hz. Magnetic memory signals were measured by the TSM-8 M testing instrument (made in Russia). The probe with a sensitivity of 1 A/m was gripped on a nonferromagnetic 3D electric scanning platform and positioned vertical to the surface of the specimen with a lift-off value of 1 mm. During testing, when a specimen was loaded to a predetermined cycle number, it was removed from the testing machine and placed on the nonmagnetic material plat-form along the South–North direction, and then the MMM signals were measured along each scanning line. A slight difference was observed in the amplitude of the signals for the three scanning lines, but the variation trend remained the same. There-fore, the relationship between the magnetic memory signals and fatigue cycles on scanning line #1 is analyzed in the following section.

10.3   Result and Discussion

10.3.1   Variations in MMM Signals Under the Fatigue Process

The distributions of the normal component of the magnetic signals $H_p(y)$ are illustrated in Fig. 10.2. These distributions correspond to different predetermined fatigue cycles. The initial magnetic signals are in the range of −60 to −40 A/m. This finding indicated that both kinds of specimens have low and stable residual magnetism after demagnetization. Figure 10.2a shows that each magnetic signal curve is almost linear with a small fluctuation at the notch location (at the testing length of 35 mm). In contrast, obvious peak-trough characteristics can be observed in the notch area of the 45 steel specimen, as shown in Fig. 10.2b. However, the curves are also almost linear along the scanning line.

Figure 10.3 shows the slope calculation of each magnetic signal curve to further quantify the relationship between MMM signals and fatigue damage of specimens. This figure shows that the slopes of both specimens sharply increased after a few loading cycles. In this stage, the magnetic domain motions caused by the stress led the specimen to enter a magnetic ordered state. The slopes went into a fluctuation stage after a few cycles. This result is attributed to random factors during the experiments, such as the loading or unloading process and clamping effect. Before fracture, visible cracks appeared on the surfaces of both the PTAW coating and 45 steel specimens.

Figure 10.4 shows the variation in the tangential component of MMM signal $H_p(x)$ with the loading cycles. The figure shows that the $H_p(x)$ values of the PTAW coating and 45 steel specimens are in the range of 10–20 A/m before loading. This finding indicates that the specimens have a low magnetic field intensity after demagnetization. The $H_p(x)$ curves of the PTAW coating are almost horizontal without

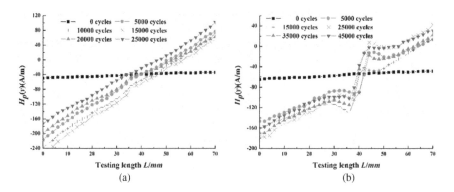

Fig. 10.2 $H_p(y)$ distributions on scanning line #1 in the fatigue process of specimens: **a** PTAW coating and **b** 45 steel

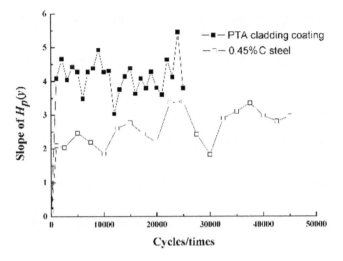

Fig. 10.3 Variation in the slope of MMM signals $H_p(y)$ with fatigue cycles

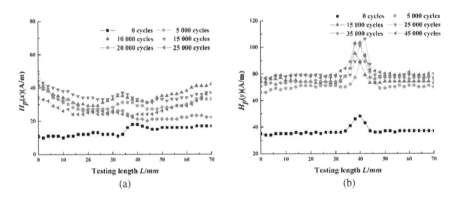

Fig. 10.4 $H_p(x)$ distributions on scanning line #1 in the fatigue process of specimens: **a** PTAW coating and **b** 45 steel

obvious peaks in the notch area, as shown in Fig. 10.4a. However, Fig. 10.4b shows the noticeable peak of the $H_p(x)$ curves of the 45 steel. The figure also shows that the intensity of $H_p(x)$ is greater than that of the cladding coating. The averages of $H_p(x)$ are calculated to investigate the difference in the surface magnetic signals between the PTAW coating and 45 steel, as shown in Fig. 10.5. The average value of $H_p(x)$ increases sharply at the initial stage and then reaches a fluctuation stage, which is similar to the variation in $H_p(y)$ shown in Fig. 10.3.

Before failure, the PTAW coating specimen underwent 25,080 fatigue cycles, and the 45 steel specimen experienced 50,183 cycles. Figure 10.6 shows the distribution of the MMM signals when the specimen fractured. Figure 10.6a shows that the $H_p(y)$ curves have an obvious peak and trough because of macrocrack. The amplitude of

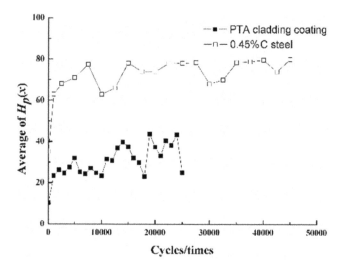

Fig. 10.5 Relationship between the average value of $H_p(x)$ and the loading cycles

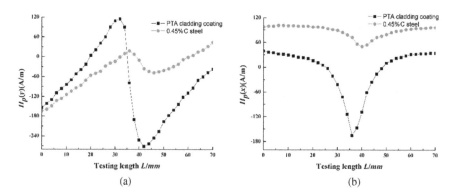

Fig. 10.6 MMM signals after the specimens fracture: **a** $H_p(y)$; and **b** $H_p(x)$

$H_p(y)$ for the PTAW coating dramatically increases in the fracture zone compared with the 45 steel specimen. The maximum value of $H_p(x)$ of the coating is also higher than that of the 45 steel, as shown in Fig. 10.6b.

10.3.2 Comparison of the Magnetic Properties from Different Material Layers

To further analyze the relationship between the MMM signals and the magnetic properties of different materials, a Lake Shores 7307 vibrating sample magnetometer

(VSM) was adopted to measure the magnetic properties of different layers at the cladding coating. The hysteresis curves are shown in Fig. 10.7. The magnetic properties of the cladding layer, bonding layer and substrate were calculated, as shown in Table 10.3. These calculations are based on the curves in Fig. 10.7. The saturation magnetization M_s of the PTAW layer is lowest among the three layers. This finding can be attributed to the fact that 12.5–14.5% of nonferromagnetic materials (Cr etc.) in the PTAW alloy powder result in the smallest magnitude of atomic magnetic moments. Moreover, the initial susceptibility χ_{in} of the PTAW layer is the smallest, whereas the coercivity H_c is the largest among the three layers. This finding was obtained because the cladding coating contains several dendrite structures and impurities that increase the dislocation density and pin sites within a ferromagnet.

The J-A model was also used to characterize the relation between the magnetization field and the stress of both the cladding coating and 45 steel. Based on the J-A model, the application of stress to a magnetic material can cause a complete reorganization of the domain structure. According to the piezomagnetic effect, the magnetization M of a ferromagnetic material under the action of stress is close to that of anhysteretic magnetization M_{an} [9]:

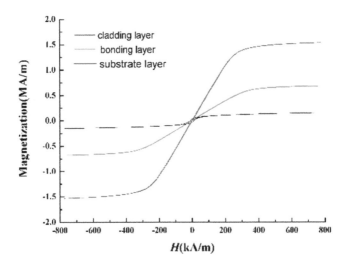

Fig. 10.7 Hysteresis loop of the PTAW coating specimen obtained from VSM testing

Table 10.3 Magnetization characteristics of different layers of the PTAW coating (SI units)

Sampling position	Saturation Magnetization M_s (MA/m)	Initial susceptibility χ_{in}	Coercivity H_c (kA/m)
Cladding layer	0.153	1.472	11.85
Bonding layer	0.680	1.956	0.589
Substrate layer	1.530	5.508	0.541

$$\frac{dM}{d\sigma} = \frac{1}{\varepsilon^2}\sigma(M_{an} - M) + c\frac{dM_{an}}{d\sigma} \tag{10.1}$$

where σ denotes the stress; $\varepsilon = (E\xi)^{1/2}$; E is the elastic modulus; and ξ indicates the hindering degree of the magnetic domain wall caused by dislocation, which could characterize the distance between M and M_{an}. When ξ is small, the distance will be small.

The effect of stress on magnetization can be equivalent to the effect of the magnetic field:

$$H_{eff} = H + \alpha M + \frac{3}{2}\frac{\sigma}{\mu_0}\left(\frac{\partial\lambda}{\partial M}\right)_\sigma \tag{10.2}$$

where H is the external magnetic field intensity and λ is the magnetostrictive coefficient.

Anhysteretic magnetization M_{an} can be represented by the Langevin equation:

$$M_{an}(H_{eff}) = M_s\left[\coth\left(\frac{H_{eff}}{a}\right) - \frac{a}{H_{eff}}\right] \approx \frac{M_s}{3a}H_{eff} \tag{10.3}$$

The computation results based on these equations are shown in Fig. 10.8. In the calculation, the model parameters that use the magnetostriction data of Kuruzar and Cullity [4] are a = 900 A m^{-1}, k = 2000 A m^{-1}, α = 1.1 × 10^{-3}, γ_{11} = 2 × 10^{-18} A^{-2} m^2, γ_{12} = 1.5 × 10^{-26} A^{-2} m^2 Pa^{-1}, γ_{21} = 2 × 10^{-30} A^{-4} m^4, γ_{22} = 5 × 10^{-39} A^{-4} m^4 Pa^{-1}, ξ = 605 Pa, and ε = 1.1 × 10^7 Pa. Figure 10.8 shows that the magnetization initially increases before it decreases. During the unloading

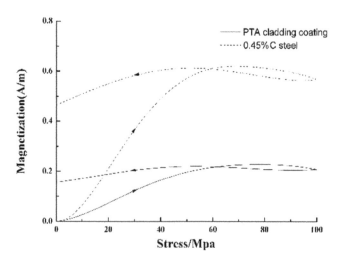

Fig. 10.8 The variation in the magnetization with stress for the PTAW coating and 45 steel specimens

process, magnetization does not change along the original path but increases at first and then decreases. The curves do not return to the original location because of the irreversible rotation of the internal magnetic domain. This finding shows that the decreasing rate of magnetization in the unloading process is lower than the increasing rate during loading. The difference between the two magnetization curves should also be considered. The amplitude of magnetization for the PTAW coating is smaller than that of 45 steel because the impurities and dendrite structure in the coating impede magnetic domain motion.

10.3.3 Microstructure Analysis

During fatigue processing, the transformation of the microstructure with accumulated cycles would lead to the variation in the magnetic signals. The microstructures of the PTAW coating under different loading cycles were observed and illustrated in Fig. 10.9 to qualitatively analyze the relationship between the magnetization and the microphase. Numerous dendrites in the remanufactured coating layer were found in the initial stage, as shown in Fig. 10.9a. These dendrites prevent magnetic domain motion during the magnetization process. This finding is consistent with the VSM test results shown in Table 10.3. As shown in Fig. 10.9b, dendrites in the microstructure are not obvious in the stable stage. This finding suggests that dislocation is the main cause of hysteresis. Fatigue damage increases slowly at this stage. Thus, the magnetic signals remain stable (the slope of $H_p(y)$ and the average of $H_p(x)$ fluctuate around a constant value), as shown in Figs. 10.3 and 10.5.

Stress was immediately released when the specimen was broken. The demagnetizing field was induced at the instant of fracture. Figure 10.6 shows the sharp increase in the MMM signal. This phenomenon can be explained in terms of the interaction

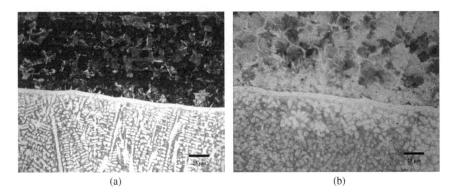

(a)   (b)

Fig. 10.9 Microstructures of the PTAW coating and substrate: **a** at the 100th cycle; **b** at the 9000th cycle

energy in the ferromagnetic material. The total free energy of the ferromagnetic material, which includes the magnetocrystalline anisotropy energy F_k, stress energy F_σ and demagnetization energy F_d, can be described by Eq. (10.4) when the temperature is far below the Curie point:

$$F = F_k + F_\sigma + F_d \tag{10.4}$$

The total energy induced by the stress should accumulate when magnetic materials are subjected to an applied stress. Stress forms domain wall motion and prevents changes in energy, which causes magnetization of the specimen. Stress energy F_σ is immediately released when the specimen is fractured. The demagnetization energy F_d increases dramatically to enable the system to recover to a balanced state. The specimen is then divided into two independent magnets. Figure 10.6 shows that the demagnetization energy of the PTAW coating is larger than that of 45 steel.

10.4   Conclusions

In this chapter, variations in the magnetic signals on the surface of the PTAW remanufactured coating specimens were investigated and compared with that of the substrate (45 steel). The slope of $H_p(y)$ and the average of $H_p(x)$ increased sharply in the few initial cycles for the coating and 45 steel specimens. These values fluctuated around a constant value. The slope of $H_p(y)$ of the PTAW coating is steeper, and the average of $H_p(x)$ is smaller than that of 45 steel because of the differences in the magnetic properties, such as the saturation magnetization, initial susceptibility, coercivity and remanence. This result was verified by VSM testing and metallographic analysis. Therefore, the slope of $H_p(y)$ and the average of $H_p(x)$ are potentially useful indicators for monitoring and evaluating the surface quality for PTAW remanufactured coatings under fatigue loading. However, this method is still under qualitative evaluation. Several issues need to be addressed, such as the quantitative detection of fatigue damage and evaluation of the surface quality of PTAW remanufactured coatings using the MMM method.

References

1. R. Veinthal, F. Sergejev, A. Zikin et al., Abrasive impact wear and surface fatigue wear behaviour of Fe-Cr-C PTA overlays. Wear **301**(1), 102–108 (2013)
2. F. Garcíavázquez, Analysis of weld bead parameters of overlay deposited on D2 steel components by plasma transferred arc (PTA) process. Mater. Sci. Forum **755**, 39–45 (2013)
3. H. Dikbas, U. Caligulu, M. Taskin et al., X-ray radiography of Ti6Al4V welded by plasma tungsten arc (PTA) welding. Mater. Prufung **55**(3), 197–202 (2013)

4. C.C. Li, L.H. Dong, H.D. Wang, et al. Metal magnetic memory technique used to predict the fatigue crack propagation behavior of 0.45% C steel. J. Magn. Magn. Mater. **405**, 150–157 (2016)
5. H.H. Huang, S.L. Jiang, R.J. Liu et al., Investigation of magnetic memory signals induced by dynamic bending load in fatigue crack propagation process of structural steel. J. Nondestruct. Eval. **33**(3), 407–412 (2014)
6. S. Bao, H.J. Lou, S. Gong, Magnetic field variation of a low-carbon steel under tensile stress. Insight-Non-Destruct. Test. Cond. Monit. **56**(5), 252–255 (2014)
7. M. Ahmad, A. Arifin, S. Abdullah, Evaluation of magnetic flux leakage signals on fatigue crack growth of mild steel. J. Mech. Eng. Sci. **9**, 1827–1834 (2016)
8. H.H. Huang, Z.C. Qian, C. Yang et al., Magnetic memory signals of ferromagnetic weldment induced by dynamic bending load. Nondestruct. Test. Eval. **32**(2), 166–184 (2017)
9. D.C. Jiles, Theory of the magnetomechanical effect. J. Phys. D Appl. Phys. **28**, 1537–1546 (1995)

Chapter 11
Detection and Evaluation of Coating Interface Damage

11.1 Introduction

Although metal magnetic memory (MMM) evaluation of heat residual stress and service fatigue damage for remanufactured coatings has been investigated in Chaps. 9 and 10, respectively, MMM evaluation is still unable to meet the requirements of holistic quality evaluation for remanufactured components. Generally, remanufactured components are composed of three parts: the coating, substrate and bonding interface. The coating/interface/substrate can be seen as a kind of metastable multimaterial system, and the interface is a weak area where the stress, crack or fracture may first initiate. The mechanical properties between the coating and substrate will differ under the effect of applied loads [1]. Thus, the stress intensity factor and the stress field distribution are complex and present oscillatory singularities at the interfacial layer [2], which results in a decrease in the bonding strength between the coating and substrate [3]. Interfacial damage will accumulate during the service process, and subsequently, microcracks will initiate, combine and propagate along the interface. Finally, fracture or peeling will appear in the coatings [4], which negatively affects the structural integrity of remanufacturing components.

MMM signals may have the potential to evaluate the fatigue damage state of the remanufactured coating interface. To accurately evaluate the fatigue damage and crack propagation length along the interface, it is necessary to establish a suitable mode. In 1995, Jiles discussed the relationship among the stress, magnetization and magnetostriction and then proposed the magnetomechanical model based on the magnetic domain pinning effect, law of approach and effective field theory [5]. This phenomenological model has been widely applied to the simulation of magnetization variation due to its advantages of simple calculation, good stability and specific physical meaning. To explain the phenomenon of spontaneous magnetization, Shi [6] and Xu [7] et al. modified the classical magnetomechanical model. The results of their theoretical calculation could accurately predict the variation in magnetic signals.

© Science Press 2021   181
H. Huang et al., *Metal Magnetic Memory Technique and Its Applications in Remanufacturing*, https://doi.org/10.1007/978-981-16-1590-0_11

Previous studies on magnetomechanical models have primarily focused on the constitutive relationship between stress and magnetization but have rarely addressed the evaluation of fatigue cracks for coating interfaces.

It is worth noting that the concept of the cohesive zone introduced from fracture mechanics is particularly useful for interfacial cracking analysis [8–10]. The origin of the cohesive zone model goes back to the experiments conducted by Dugdale [11], and later, it was embedded into numerical simulations. Compared with traditional fracture mechanics analysis tools, such as the stress intensity factor K and J integration, the cohesive zone model can make accurate predictions for the entire fracture process (including crack initiation and propagation up to the onset of catastrophic failure). Furthermore, the cohesive zone model can simulate the fatigue behavior of materials that violate small-scale yielding assumptions at the crack tip. The stress singularity can also be neglected in this model by establishing the traction displacement relationship for the crack tip. This approach is independent of geometries and loading configurations and is particularly capable of performing cracking analysis for nonlinear structures. Based on these advantages, Roe et al. [10] employed the damage evolution law in the cohesive zone model. The model was later successfully extended to other applications, such as fatigue crack propagation analysis in brazed joints [8], solder joints [9], ductile metallic layers [12], single crystal superalloys [13] and adhesive bonds [14].

One key issue in the evaluation of interfacial fatigue cracks in this chapter is to combine the cohesive zone concept with the magnetomechanical model to establish the fatigue cohesive zone-magnetomechanical coupling model for MMM testing. The interfacial crack initiation time and propagation behavior are then analyzed and predicted under three-point bending fatigue loads. Furthermore, a new magnetic characteristic indicator is proposed to characterize the crack propagation rate. Finally, the theoretical model is verified by reference experiments, which lay a foundation for MMM testing in remanufactured coating interfaces. The rest of this chapter is structured as follows. In Sect. 11.2, the theoretical framework of the whole research is established, including the fatigue cohesive model, the magnetomechanical model and the numerical algorithm of the coupling model, and the calculation formula of the magnetic field intensity is given. The case analysis for the theoretical model is described in detail in Sect. 11.3. The interfacial crack initiation and the interfacial crack propagation behavior of the remanufactured coating are predicted by using this model. To verify the results of the theoretical analysis, the MMM signals of the remanufactured coating interface under three point bending fatigue loads are collected and analyzed by the experiment in Sect. 11.4.

11.2 Theoretical Framework

11.2.1 Fatigue Cohesive Zone Model

A cohesive zone is assumed in front of the crack tip along the interface of potential crack propagation, as shown in Fig. 11.1a. Since the attraction between atoms is closely related to their distance, the traction σ between the coating and substrate in the cohesive zone can be considered a function of the displacement u between the top surface and bottom surface of the interfacial crack. The behavior of the materials at each cohesive element is controlled by the traction-displacement constitutive relationship, as shown in Fig. 11.1b. The static cohesive law is dependent on three key parameters, including the critical traction σ_0, critical displacement u_0 and maximum displacement u_f:

$$\sigma = \begin{cases} \frac{\sigma_0}{u_0} u & u \leq u_0 \\ \sigma_0 \frac{u_f - u}{u_f - u_0} & u > u_0 \end{cases} \tag{11.1}$$

At the OA stage in Fig. 11.1b, the traction increases with increasing displacement. When the displacement exceeds the critical value u_0, irreversible material degradation occurs, and the traction decreases along line AB. If the displacement reaches the maximum value u_f, the cohesive element cannot bear any more traction, and a crack appears. To investigate the initiation and propagation of fatigue cracks, the cyclic damage evolution law should be introduced [13]:

$$\Delta D_c = A(1 - D_c)^m \left(\frac{\sigma}{1 - D_c} - \sigma_0 \right)^n \|\Delta u\| \tag{11.2}$$

where A, m and n are the parameters that control the damage evolution. σ_0 is the critical traction under which the damage does not accumulate. Δu is the difference in the displacement of the cohesive zone. D_c and ΔD_c represent the damage accumulation degree and its variation, respectively. Analogously, the cohesive element will also be

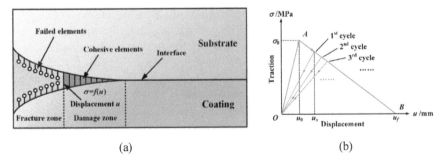

(a)   (b)

Fig. 11.1 Fatigue cohesive zone model: **a** cohesive zone; and **b** traction-displacement relationship

degraded when the displacement exceeds the critical value u_0 under fatigue loading. The degraded stiffness can be described as $k = k_0(1 - D_c)$, where k_0 is the initial stiffness. The fatigue traction-displacement constitutive relationship can be given as [9]:

$$\sigma = k_0(1 - D_c)u \tag{11.3}$$

In Eq. (11.3), it is assumed that each unloading path always follows the original loading path and that the damage degree does not evolve during the unloading process, as shown in Fig. 11.1b. With increasing loading cycle number N, the damage degree D_c will increase while the stiffness k and the traction σ decrease. When the damage degree D_c accumulates to the value of 0.99, the stiffness k or the traction σ approaches zero, and the failed cohesive element can be removed. Furthermore, the fatigue traction-displacement paths vary with the initial displacements u_x. The number of cycles to failure for cohesive elements decreases with increasing initial displacement. It should be noted that when the initial displacement is below the critical value u_0, the material damage cannot accumulate, and the lifetime can be considered to be infinite.

11.2.2 Magnetomechanical Model

The interface between the coating and substrate is a kind of ferromagnetic material, where spontaneous magnetization will be induced under the applied load. The relationship between the interfacial stress and magnetization can be determined by the Jiles magnetomechanical model. For a single interfacial layer with a certain thickness, it can be assumed that the block material is isotropic. Based on the magnetomechanical model and the theoretical equations established in Chap. 2, the relationship between the magnetization M and stress σ can also be described as:

$$\frac{dM}{d\sigma} = \frac{\sigma}{E_a\xi}(M_{an} - M) + c\frac{dM_{an}}{d\sigma} \tag{11.4}$$

where ξ is a coefficient related to the energy density, E_a is the elastic modulus for the interface and c represents the flexibility of the magnetic domain walls. In this chapter, we propose that the stress in the magnetomechanical model is equivalent to the traction in the cohesive zone model. Therefore, considering Eqs. (11.2), (11.3) and (11.4), the variations in the magnetization M during the cyclic loading-unloading process in each cohesive element can be obtained.

11.2.3 Numerical Algorithm of the Coupling Model

The calculation process that couples the fatigue cohesive zone model and the magnetomechanical model is shown in Fig. 11.2. First, relevant parameters are input into the cohesive zone model. The traction $\sigma(N)$, displacement $u(N)$ and damage accumulation $D_c(N + 1)$ under each loading cycle can then be obtained. If the damage accumulation $D_c(N + 1)$ exceeds the critical threshold (the value is set to 0.99 in this chapter), the failed cohesive element can be removed. Otherwise, the next cycle of calculation is conducted. During the calculation process of the cohesive zone model, each traction $\sigma(N)$ should be input into the magnetomechanical model to control the corresponding magnetization history. The variations in the magnetization during the loading and unloading processes are calculated. Finally, the number of cycles to crack initiation, the cyclic traction-displacement relationship, the damage evolution process and the cyclic traction-magnetization curves can be obtained. Based on this, crack initiation and propagation can be predicted.

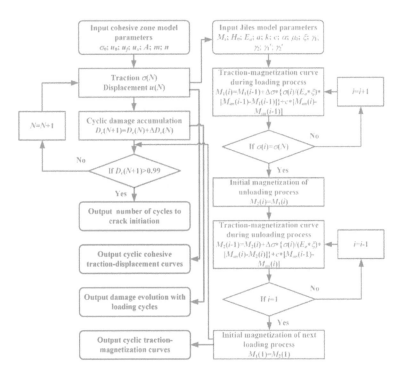

Fig. 11.2 Calculation process of the theoretical model

11.2.4 *Calculation of the Magnetic Field Intensity*

In magnetic nondestructive testing applications, the damage is evaluated by a measurable magnetic field intensity rather than magnetization. To predict interfacial crack initiation based on the magnetic field intensity measured from the surface of ferromagnetic materials, the cohesive element can be considered a magnetic dipole. The magnetic vector potential A far away from the magnetic dipole can be described as:

$$A = \frac{\mu_0}{4\pi} \frac{m \times r}{r^3} \tag{11.5}$$

where m is the magnetic dipole moment and r is the radius vector. The magnetic induction intensity B in space can be given as follows:

$$B = \nabla \times A = \frac{\mu_0}{4\pi r^3}\left[\frac{3}{r^2}(m \cdot r)r - m\right] \tag{11.6}$$

When the volume of the magnetic dipole is $V = 4\pi a^3/3$, the magnetic dipole moment is $m = VM$, where M is the magnetization. The relationship between the magnetic field and magnetization can be established, and the magnetic field intensity H in air is obtained:

$$H = \frac{B}{\mu_0} = \frac{1}{3}\left(\frac{a}{r}\right)^3\left[\frac{3}{r^2}(M \cdot r)r - M\right] \ (r > a) \tag{11.7}$$

11.3 Case Analysis for the Theoretical Model

11.3.1 *Finite Element Model Setup*

The specific geometry and the loading configuration should be determined first to validate the theoretical model in magnetic nondestructive testing. The typical three-point bending fatigue test and its sinusoidal waveform loading spectrum are selected as shown in Fig. 11.3a and b, respectively. The maximum amplitude of the loading spectrum is 0.25 mm with a stress ratio of 0 and a frequency of 1 Hz, which generates the maximum load at 10.4 kN.

The nickel-based coating is deposited on the 45 steel substrate, and an interface will be naturally formed. From the micro perspective, the material properties of the metallurgical bonding layer and its adjacent heat-affected zone are different from those of the coating or substrate due to their various microstructures. Therefore, the bonding layer and the heat-affected zone can be considered an interface layer with a certain thickness. The material properties of the coating, interface and substrate are listed in Table 11.1 based on Refs. [15, 16].

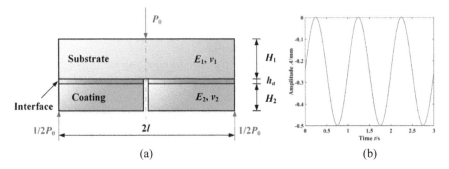

Fig. 11.3 Three-point bending fatigue test: **a** schematic diagram and **b** loading spectrum

Material properties	Coating	Interface	Substrate
Tensile strength (MPa)	800–900	400–1200	400–600
Hardness (HV)	700	200–500	<200
Elastic modulus (GPa)	280	260	210
Poisson's ratio	0.22	0.24	0.30
Saturation magnetization (A/m)	0.2×10^6	1.7×10^6	1.5×10^6
Initial susceptibility	1.5	2.0	5.5
Coercivity (A/m)	11.9×10^3	0.6×10^3	0.5×10^3

Table 11.1 Material properties of the coating, interface and substrate [15, 16]

A sharp precrack is machined in the middle position of the coating layer to avoid sudden brittle fracture during the loading process. The evolution analysis of crack length a can be limited to the right-hand half-portion of the interface due to the axial symmetry of the geometry and loading configuration. In this chapter, the length of the supporting span is denoted by $2l = 80$ mm. The thicknesses of the substrate and coating are denoted by $H_1 = 10$ mm and $H_2 = 5$ mm, respectively, and the thickness of the interface in between is $h_a = 0.5$ mm.

Based on this, the entity model of the specimen is established by the finite element software ABAQUS, as shown in Fig. 11.4. The local coordinate system is fixed with its origin at the middle position of the interface. The coating, interface and substrate models are all generated by a quadrilateral mesh. The elements of the coating and substrate are CPS4R with a mesh size of 1 mm. Considering that the stress calculation is only a linear static stress solving process, the mesh quality of the precrack tip zone is refined with a mesh size of 0.25 mm to guarantee accuracy. The interface is meshed by a single layer of cohesive elements COH2D4 with a thickness of 0.5 mm. There are 5693 nodes and 4869 elements in total after meshing, where the number of cohesive elements is 478.

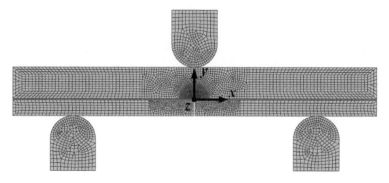

Fig. 11.4 Model setup of the finite element simulation for three-point bending loading

11.3.2   Finite Element Simulation Results

At the first loading cycle with an amplitude of −0.5 mm, the pressure head moves down 0.5 mm, and the stress distribution is shown in Fig. 11.5. Stress discontinuity occurs at the interface due to the different material properties of the coating, interface and substrate. The high interfacial stress is concentrated in the zone within 10 mm of the precrack. The maximum interfacial stress appears at the precrack position. The stress intensity factor is adopted here as a critical parameter to describe the stress field around the crack tip, whose distribution at different crack lengths along the interface is shown in Fig. 11.6. The maximum stress intensity factor also appears at the precrack position. When the interfacial crack initiates, the stress intensity factor at the new crack tip shows a decreasing trend. As a result, the crack is first initiated from the precrack where the stress intensity factor is larger and then propagates along the interface.

The displacement u in the cohesive zone model represents the relative displacement of the cohesive element, as shown in Fig. 11.7. Under the effect of the first loading cycle, the distribution of the initial displacements u_x along the interface

Fig. 11.5 von Mises stress contour plot of three-point bending loading

Fig. 11.6 Distribution of
stress intensity factors K_I at
different crack lengths along
the interface

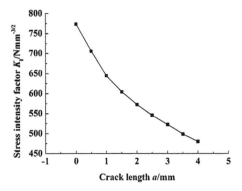

Fig. 11.7 Displacement of
the cohesive finite elements
in the simulation

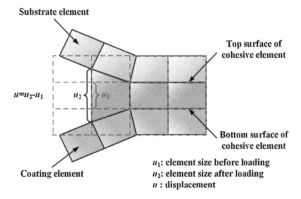

within the high stress concentration zone is plotted in Fig. 11.8. The initial displacement of the cohesive element decreases with increasing distance from the precrack.
When the distance exceeds 4 mm, the initial displacement remains stable.

Fig. 11.8 Distribution of
initial displacements u_x of
the cohesive elements along
the interface

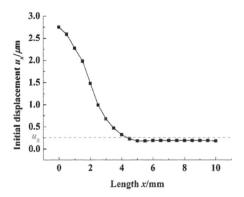

11.3.3 *Prediction of Interfacial Crack Initiation*

Based on the aforementioned simulation results, the interfacial fatigue crack first initiates at the precrack position where the initial displacement reaches the maximum at $u_x = 2.75 \times 10^{-3}$ mm. The remaining parameters of the fatigue cohesive zone model are set to $\sigma_0 = 400$ MPa, $u_0 = 2.5 \times 10^{-4}$ mm, $u_f = 1.6 \times 10^{-2}$ mm, $A = 20$, $m = 2$ and $n = 1/4$. Based on numerical calculations, the fatigue traction-displacement paths can be given in Fig. 11.9. The red dotted line represents the static traction-displacement relationship. The interfacial damage at the precrack will accumulate after one cycle because the initial displacement u_x exceeds the critical displacement u_0. Both the stiffness and traction decrease with increasing number of cycles. During the fatigue process, the damage evolution is shown in Fig. 11.10. It can be found that the damage degree D_c increases, while its variation rate decreases, with increasing number of cycles. When the cycle number reaches $N = 354$, the damage degree is $D_c = 0.99$, and the stiffness and traction approach zero. It can be considered that interfacial crack initiation occurs at this moment.

Fig. 11.9 Interfacial fatigue traction-displacement relationship at the precrack

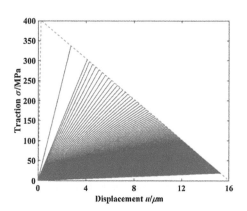

Fig. 11.10 Interfacial damage evolution process at the precrack

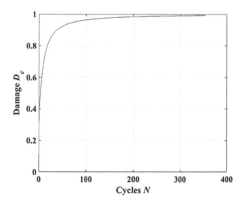

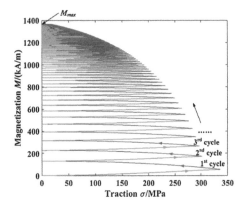

Fig. 11.11 Variations in interfacial magnetization during the cyclic loading-unloading process at the precrack

To predict the crack initiation time, the parameter values in the magnetomechanical model are set to $M_s = 1.7 \times 10^6$ A/m, $a = 1000$ A/m, $\alpha = 0.008$, $c = 0.1$, $\mu_0 = 4\pi \times 10^{-7}$ H/m, $H_0 = 40$ A/m, $E_a = 260$ GPa, $\xi = 5 \times 10^6$ Pa, $\gamma_1(0) = 2 \times 10^{-18}$ A^{-2} m^2, $\gamma_1'(0) = -1 \times 10^{-26}$ A^{-2} m^2 Pa^{-1}, $\gamma_2(0) = 1 \times 10^{-30}$ A^{-4} m^4, and $\gamma_2'(0) = -5 \times 10^{-39}$ A^{-4} m^4 Pa^{-1}. The variations in magnetization during the cyclic loading-unloading process can be given in Fig. 11.11. The red and blue curves represent the magnetization paths under the loading and unloading processes, respectively. It can also be found that the traction decreases with increasing number of cycles. In addition, the magnetization degree increases, while its accumulation rate decreases with increasing number of cycles. When the cycle number reaches $N = 354$, crack initiation occurs, and the corresponding magnetization accumulates to the maximum $M_{max} = 1362.6$ kA/m. The lift-off value of the probe is set to $r = 5$ mm and the size of the magnetic dipole is set to $a = 0.32$ mm. Based on Eq. (11.7), the magnetic field intensity increases to the threshold $H_{max} = 238.1$ A/m when the interfacial crack initiates. This indicates that interfacial crack initiation can be predicted by measuring the spontaneous magnetic flux leakage signals at the precrack.

11.3.4 Prediction of the Interfacial Crack Propagation Behavior

The interfacial crack propagation length a is deduced by calculating the number of cycles to element failure N. When the crack propagates to a certain length, the corresponding loading cycle number can then be determined. Based on the traction-separation relationship in Fig. 11.1b, various initial displacements can lead to different numbers of cycles to failure. Therefore, some specified interfacial cohesive elements with different initial displacements in Fig. 11.8 are chosen, as listed in Table 11.2. After putting these initial displacements u_x into the theoretical model, the corresponding number of cycles to failure N, the maximum magnetization M_{max} and the maximum magnetic field intensity H_{max} can be calculated. The interfacial

Table 11.2 Theoretical calculation results

Position of the specified elements x/mm	0	0.5	1.0	1.5	2.0	2.5	3.0	3.5	4.0
Initial displacement u_x/μm	2.75	2.59	2.27	1.98	1.48	0.99	0.68	0.47	0.32
Number of cycles to failure N	354	372	418	473	629	978	1600	3003	9119
Maximum magnetization M_{max} (kA/m)	1362.6	1420.7	1508.5	1552.7	1572.7	1573.6	1574.3	1575.2	1575.7
Maximum magnetic field intensity H_{max} (A/m)	238.13	248.29	263.63	271.35	274.85	275.01	275.13	275.29	275.37
Crack propagation length a/mm	0	0.5	1.0	1.5	2.0	2.5	3.0	3.5	4.0

Fig. 11.12 Relationship among the interfacial crack propagation length a, maximum magnetic field intensity H_{max} and fatigue cycle numbers N

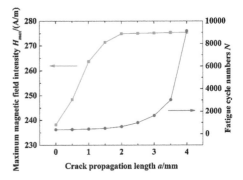

crack tip will also propagate to the position of the failed cohesive element. Based on the data in Table 11.2, the relationship among a, H_{max} and N can be obtained as shown in Fig. 11.12.

Generally, the propagation rate of mode I cracks increases with increasing fatigue cycle numbers. However, the propagation behavior of interfacial cracks shows an inverse association under the three point bending fatigue loads, as shown in Fig. 11.12. The crack propagation rate decreases with increasing number of loading cycles. One possible explanation is that both the stress intensity factor and its variation rate decrease with increasing crack length, as shown in Fig. 11.6. This indicates that more fatigue cycle numbers are required to break the cohesive element before the crack propagates to a new position. At the initial stage of crack propagation, the cohesive element of the crack tip close to the precrack position has a larger initial displacement, as shown in Fig. 11.8. Based on the fatigue cohesive zone model, the number of cycles to failure for cohesive elements decreases with increasing initial displacement. This means that the interfacial fatigue crack can easily initiate at the precrack position where the initial displacement reaches a maximum of $u_x = 2.75 \times 10^{-3}$ mm. When the crack propagates to a length of 4 mm, the initial displacements are very close to the critical value u_0, and more failure cycle numbers are required to break the element. Therefore, the crack propagation rate decreases for higher loading cycles. Analogously, the maximum magnetic field intensity increases, while its accumulation rate decreases with increasing crack length. When the maximum magnetic field intensity remains stable, it can be considered that the material at the crack tip reaches the saturation magnetization degree. These results indicate that a quantitative relationship may be further found from Fig. 11.12.

To quantitatively characterize the crack propagation rate da/dN, a new magnetic characteristic indicator, the magnetic field accumulation rate dH_{max}/dN, is proposed, as shown in Fig. 11.13. At the fatigue crack stable propagation stage (i.e., Region II), the crack propagation rate da/dN increases linearly with increasing magnetic field accumulation rate dH_{max}/dN, and their fitting equation can be given as follows:

$$\lg\left(\frac{da}{dN}\right) = \lg C_1 + m_1 \cdot \lg\left(\frac{dH_{\max}}{dN}\right) \tag{11.8}$$

Fig. 11.13 Relationship between the magnetic field accumulation rate dH_{max}/dN and crack propagation rate da/dN

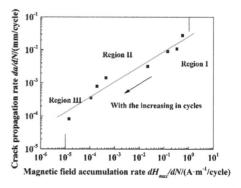

where $\lg C_1$ and m_1 are constant values related to the material properties and loading configurations. In this case, $\lg C_1$ is -1.592, m_1 is 0.461 and R_1^2 (adjusted deviate square) is 0.937. The level of R_1^2 is high enough to guarantee the fitting goodness. At the crack initiation stage (i.e., Region I), the magnetic field accumulation rate and the crack propagation rate are both rapid. However, at the ultimate stage (i.e., Region III), the magnetic field accumulation rate and the crack propagation rate are both slow. As a result, the interfacial crack propagation rate can be effectively evaluated by the magnetic field accumulation rate, which has the potential to be applied to magnetic nondestructive testing.

11.4 Experimental Verification

11.4.1 MMM Measurement Method

The nickel-based wear-resistant coating was deposited on the 45 steel substrate by a typical surface engineering technique called plasma transferred arc welding. The chemical compositions of the nickel-based self-fluxing alloy powder (Ni60) and 45 steel are shown in Table 9.1 and Table 5.1, respectively. A specimen with the same dimensions of the coating/interface/substrate sandwich structure as that listed in the aforementioned case analysis was prepared. Weak residual stress or residual magnetization will inevitably be generated during specimen preparation. To eliminate the negative effect of these interferences on the detection results, the specimen was annealed at 550 °C for 5 h in a vacuum furnace and demagnetized before testing. Then, the three-point bending fatigue test was carried out on an SDS-100 hydraulic testing machine, as shown in Fig. 11.14. The dynamic loading spectrum mentioned above in Fig. 11.3b was obtained. When loaded to the preset cycle numbers of $N = 354, 372, 418, 473, 629, 978, 1600, 3003$ and 9119 (corresponding to the theoretical crack lengths of $a = 0, 0.5, 1.0, 1.5, 2.0, 2.5, 3.0, 3.5$ and 4.0 mm as listed in Table 11.2), the specimen was unloaded and taken from the supporting seats.

Fig. 11.14 Three-point
bending fatigue test

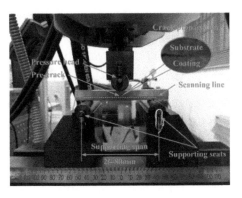

During the detection process, the specimen was placed on a nonmagnetic material platform along the South-North direction, as shown in Fig. 11.15a. The surface of the specimen with an interfacial scanning line should be kept upward. The magnetic signals were measured along this scanning line by an MMM TSC-2M-8 device. The probe with a 1 A/m resolution based on the Hall sensor was gripped on a nonferromagnetic scanning device and was placed vertical to the surface of the specimen with a lift-off value of 5 mm and horizontal movement speed of 8 mm/s. The probe can simultaneously collect the tangential components $H_p(x)$ and normal components $H_p(y)$ of the magnetic signal with a scanning increment of 1 mm. Finally, the interfacial cracks were observed by a JXD-250B optical reading microscope. The crack propagation path may not be exactly along the interface. We considered the horizontal distance between the farthest crack tip and the precrack position as the interfacial propagation length a, as shown in Fig. 11.15b. To evaluate the testing uncertainty, the magnetic signals and the crack propagation lengths under different cycle numbers were measured three times.

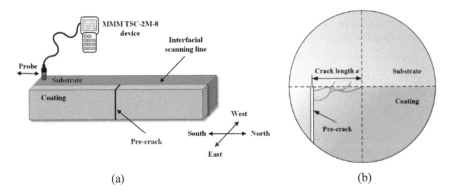

(a)   (b)

Fig. 11.15 Schematic diagram of the measurement: **a** magnetic signals and **b** crack propagation length

11.4.2   MMM Signal Analysis

Under different loading cycles, the distributions of the tangential components $H_p(x)$ and normal components $H_p(y)$ of the magnetic signal along the interface are given in Fig. 11.16. The $H_p(x)$ curves form trough features at the precrack position, and the trough values increase with increasing number of cycles. The $H_p(y)$ curves show peak-trough features, and the polarity changes near the precrack position. The peak-trough values also increase with increasing number of cycles. Based on the collected tangential and normal components of the magnetic signals, the total magnetic signals H can be defined as follows:

$$H = \sqrt{H_p^2(x) + H_p^2(y)} \tag{11.9}$$

The distribution of the total magnetic signals H along the interface is plotted in Fig. 11.17. All the curve peaks under different loading cycles are concentrated near the precrack position. Due to the limitation of resolution and accuracy in detection devices, it is difficult to distinguish the crack tip or crack length by the zero-crossing,

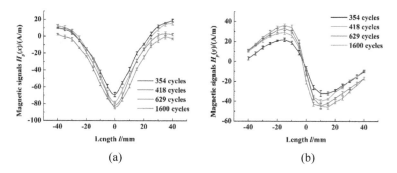

(a)   (b)

Fig. 11.16 Distributions of the magnetic signals along the interface: **a** tangential components $H_p(x)$; **b** normal components $H_p(y)$

Fig. 11.17 Distribution of the total magnetic signals H along the interface

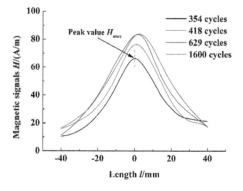

trough or peak positions in magnetic signals. Fortunately, the total magnetic field peak value H_{max} varies with the number of loading cycles, which indicates that it may have the potential to characterize fatigue cracks.

The relationship among the interfacial crack propagation length a, total magnetic field peak value H_{max} and fatigue cycle number N is given in Fig. 11.18. The interfacial crack will initiate when H_{max} reaches 69.4 A/m. Compared with Fig. 11.12, it can be found that the variation trends of the experimental results are consistent with those in the theoretical calculation. The interfacial crack length increases, while its propagation rate decreases with increasing number of cycles. Similarly, the peak value of the total magnetic field H_{max} increases, while its accumulation rate decreases with increasing crack length. However, the shapes of the curves are different because the experimental results can be affected by many factors (such as the lift-off effect of the probe, neglecting the magnetic signal component along the interfacial thickness direction and the influence of a stray environmental magnetic field). Therefore, some data fluctuations appear in Fig. 11.18.

Analogous to the theoretical analysis process, the relationship between the magnetic field accumulation rate dH_{max}/dN and crack propagation rate da/dN is shown in Fig. 11.19. It can also be found that the measured magnetic field accumulation rate dH_{max}/dN increases linearly with increasing crack propagation rate da/dN at Region II, and the fitting equation is listed as follows:

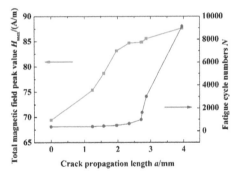

Fig. 11.18 Relationship among the interfacial crack propagation length a, total magnetic field peak value H_{max} and fatigue cycle numbers N

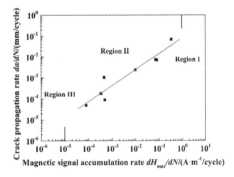

Fig. 11.19 Relationship between the magnetic field accumulation rate dH_{max}/dN and crack propagation rate da/dN

$$\lg\left(\frac{da}{dN}\right) = \lg C_2 + m_2 \cdot \lg\left(\frac{dH_{max}}{dN}\right) \tag{11.10}$$

where $\lg C_2$ and m_2 are also constant values that are dependent on the material properties and loading configurations. In this experiment, $\lg C_2$ is -1.092, m_2 is 0.761 and R_2^2 (adjusted deviate square) is 0.887, which can also guarantee the goodness-of-fit. Although the constant parameter values of the fitting equation in the experiment are different from those in the theoretical analysis, their linear increasing trend and order of magnitude are the same. Therefore, the variation trend may have important reference values for interfacial fatigue crack evaluation in coatings.

11.4.3   Interfacial Crack Observation

From the perspective of micromorphology, there is a certain distance between the precrack tip and the interface due to the machining accuracy limitation of wire cutting. Under the effect of three point bending loads, the precrack tip zone at the coating side is mainly subjected to tensile stress, as shown in Fig. 11.20a. When the cycle number reaches $N = 354$, the crack tip initiates from the precrack and immediately propagates perpendicular to the coating/substrate interface under the effect of tensile stress. The crack tip cannot cross the interface because the substrate is mainly subjected to compressive stress. The bonding stress of the interface is generally higher than the tensile stress of the coating or substrate. Therefore, the crack propagation path may not be located exactly along the interface. When the cycle number reaches $N = 372$, the crack tip propagates to the position shown in Fig. 11.20b. This indicates that the interfacial crack propagation rate is rapid at the crack initiation stage (i.e., Region I). After entering Region II, the crack propagation rate remains stable. The crack morphology at a cycle number of $N = 629$ can be found in Fig. 11.20c. During this stage, the main crack always propagates along the direction of the interface. In addition, a certain number of secondary cracks occur at the side of the coating layer due to its fragility. When the cycle number reaches $N = 9119$, the crack morphology can be seen in Fig. 11.20d. Although the width of the crack is widened, the position of the crack tip is hardly changed compared with that in Fig. 11.20c. This indicates that the crack propagation rate becomes slow and the crack propagates to the ultimate length at Region III. The observation results of the crack propagation behavior are consistent with the theoretical predictions, which can verify the fatigue cohesive zone-magnetomechanical coupling model.

In order to reveal the physical mechanism underlying generation and variation of interfacial spontaneous magnetization, the magnetic domain movement of interfacial pre-crack tip at different fatigue cycles were analyzed. At the initial stage, the random orientation of magnetic domains including 90° domain, 180° domain, and other complex shape domains. Therefore, the magnetic moments in domains offset each other and the overall zone displays weak magnetization. After the fatigue

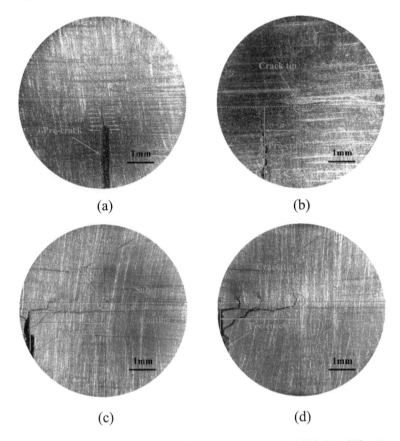

(a)   (b)

(c)   (d)

Fig. 11.20 Crack morphology under different loading cycles: **a** $N = 354$; **b** $N = 372$; **c** $N = 629$; and **d** $N = 9119$; magnification $\times 30$

cycle number reaching $N = 354$, the magnetic domains will overcome the magnetocrystalline anisotropy energy and rotate along the direction of easy magnetization. In order to minimize total magnetic energy, the magnetic domains with the same orientation will merge and collect into one domain. Due to the magnetomechanical effect, the magnetic domain wall movement and reorientation will lead to the change of magnetization in ferromagnetic materials. The Eq. (11.4) also shows that the magnetization varies under the coupling effect of the traction σ and the domain wall parameter c. After the specimen undergoes $N = 629$ loading cycles, the large dislocation density and obvious pinning effect block the movement of magnetic domains. The cohesive element materials at crack tip approach the magnetization saturation state and the magnetic field strength tends to stabilize. During the process of crack initiation and propagation, the movement of magnetic domains corresponds to the variation of the interfacial total magnetic signals H shown in Fig. 11.17.

11.5   Conclusions

In this chapter, the concept of the cohesive zone and the phenomenon of spontaneous magnetization are introduced, and the fatigue cohesive zone-magnetomechanical coupling model for MMM testing is established. Based on this, the cyclic traction-displacement relationship, the damage evolution process and the cyclic traction-magnetization curves for interfacial crack initiation are analyzed under three point bending fatigue loads with the maximum amplitude at 0.25 mm. The calculation results show that crack initiation occurs when the loading cycle number reaches 354, and the corresponding magnetic field intensity increases to the threshold H_{max} = 238.1 A/m. Furthermore, a new magnetic characteristic indicator dH_{max}/dN is proposed to characterize the crack propagation rate. The magnetic field accumulation rate dH_{max}/dN is linearly related to the crack propagation rate at the stable crack propagation stage. The crack propagation rate decreases with increasing number of loading cycles. To verify the theoretical model, the MMM signals along the interface are collected and analyzed under the three-point bending fatigue test. The experimental result is consistent with the theoretical calculation. This indicates that the fatigue cohesive zone-magnetomechanical coupling model for the coating/substrate interface established in this chapter can support MMM testing. In particular, the model can be applied to the prediction of interfacial crack initiation and propagation in remanufactured coatings.

References

1. P. Potnis, J. Holtzinger, D. Das et al., Study of fracture behaviour of bond coats on nickel superalloy by three-point bending of microbeams. Surf. Coat. Technol. **204**, 586–592 (2009)
2. X.S. Wang, C.K. Yan, Y. Li et al., SEM in-situ investigation on failure of nanometallic film/substrate structures under three-point bending loading. Int. J. Fract. **151**(2), 269–279 (2008)
3. L. Xu, H.J. Cao, H.L. Liu et al., Study on laser cladding remanufacturing process with FeCrNiCu alloy powder for thin-wall impeller blade. Int. J. Adv. Manuf. Technol. **90**, 1383–1392 (2017)
4. X.L. Fan, R. Xu, W.X. Zhang et al., Effect of periodic surface cracks on the interfacial fracture of thermal barrier coating system. Appl. Surf. Sci. **258**, 9816–9823 (2012)
5. D.C. Jiles, Theory of the magnetomechanical effect. J. Phys. D Appl. Phys. **28**(8), 1537–1546 (1995)
6. P.P. Shi, K. Jin, X.J. Zheng, A general nonlinear magnetomechanical model for ferromagnetic materials under a constant weak magnetic field. J. Appl. Phys. **119**, 145103-1-8 (2016)
7. M.X. Xu, M.Q. Xu, J.W. Li et al., Using modified J-A model in MMM detection at elastic stress stage. Nondestruct. Test. Eval. **27**(2), 121–138 (2012)
8. M.K. Ghovanlou, H. Jahed, A. Khajepour, Cohesive zone modeling of fatigue crack growth in brazed joints. Eng. Fract. Mech. **120**, 43–59 (2014)
9. A.A. Baqi, P.J.G. Schreurs, M.G.D. Geers, Fatigue damage modeling in solder interconnects using a cohesive zone approach. Int. J. Solids Struct. **42**, 927–942 (2005)
10. K.L. Roe, T. Siegmund, An irreversible cohesive zone model for interface fatigue crack growth simulation. Eng. Fract. Mech. **70**, 209–232 (2003)
11. D.S. Dugdale, Yielding of steel plates containing slits. J. Mech. Phys. Solids **8**, 100–104 (1960)

12. B. Wang, T. Siegmund, A numerical analysis of constraint effects in fatigue crack growth by use of an irreversible cohesive zone model. Int. J. Fatigue **132**, 175–196 (2005)
13. J.L. Bouvard, J.L. Chahoche, F. Gallerneau, A cohesive zone model for fatigue and creep-fatigue crack growth in single crystal super alloys. Int. J. Fatigue **31**, 868–879 (2009)
14. H. Khoramishad, A.D. Crocombe, K.B. Katnam et al., Predicting fatigue damage in adhesively bonded joints using a cohesive zone model. Int. J. Fatigue **32**, 1146–1158 (2010)
15. H.H. Huang, G. Han, C. Yang et al., Stress evaluation of plasma sprayed cladding layer based on metal magnetic memory testing technology. J. Mech. Eng. **52**(20), 16–22 (2016)
16. Z.C. Qian, H.H. Huang, W.J. Liu et al., Magnetomechanical model for coating/substrate interface and its application in interfacial crack propagation length characterization. J. Appl. Phys. **124**, 203904-1-9 (2018)

Part IV
Engineering Applications
in Remanufacturing

Chapter 12
Detection of Damage of the Waste Drive Axle Housing and Hydraulic Cylinder

12.1 Introduction

At present, the abovementioned metal magnetic memory (MMM) studies in Parts 2 and 3 mainly focused on standard specimens rather than real remanufacturing components. This is because the inconsistency of the size, shape, material and other factors of the remanufacturing components may lead to the dispersion of MMM testing results. Therefore, the fatigue test specimens with middle notches have mostly been used in previous studies based on national or international standards. For example, Li et al. [1] conducted tensile fatigue tests on 45 steel central crack tensile (CCT) specimens, and the variations in the MMM signals during fatigue testing were studied. Qian et al. [2] used a three-point bending fatigue specimen with a U-shaped notch to analyze MMM signals. Ni et al. [3] established the relationship between MMM signals and damage degree for V-notch specimens during crack initiation.

However, the laboratory testing results have not taken into account many practical influencing factors, so that the effect of MMM application in remanufacturing engineering is doubtful, and relevant research requires further verification. To improve the practicability of MMM testing, in this chapter, we use the abovementioned research results to evaluate the damage degree and remanufacturability of the retired drive axle housing and hydraulic cylinder.

The driving axle of a vehicle, which is at the end of the transmission system, not only distributes torque to the wheels but also experiences the vibration and impact from roads. In addition, defects such as pores, inclusions or cracks inevitably occur at driving axle housing during the fabrication and welding process. Therefore, the stress concentration, plastic deformation, cracks and even fractures can easily appear at the material discontinuities or welding zones under the effect of applied loads. While the vehicle is in motion, fatigue damage often occurs at the ends of the support of the driving axle housing, as shown in Fig. 12.1a.

Compared with the driving axle housing, the loading situation of the hydraulic cylinder is more complex, the service environment is tougher and the damage zone is more dispersed, as shown in Fig. 12.1b. The stress concentration on the inside and

© Science Press 2021
H. Huang et al., *Metal Magnetic Memory Technique and Its Applications in Remanufacturing*, https://doi.org/10.1007/978-981-16-1590-0_12

(a)   (b)

Fig. 12.1 Retired components to be remanufactured: **a** drive axle housing; and **b** hydraulic cylinders

outside surfaces will lead to cracks or spalling of the coating, which greatly affects the fatigue life. Therefore, it is challenging to determine the damage location and degree of the hydraulic cylinders.

In this chapter, the distribution of MMM signals in the high risk zone of drive axle housing is extracted, and the variation in the magnetic signal characteristic value with fatigue cycles and the deformation degree is discussed in Sect. 12.2. Then, based on the damage feature of the hydraulic cylinder, the damage threshold determination method is established to evaluate the remanufacturability in Sect. 12.3. The research results can provide theoretical guidance and an experimental basis for realizing the industrialization of MMM testing in remanufacturing.

12.2 Application of MMM in the Evaluation of Fatigue Damage of the Drive Axle Housing

12.2.1 Relation Between MMM Signals and Fatigue Cycles

The drive axle housing is always affected by the impact loads generated between the wheels and ground, which may cause deformation or even cracks at the ends of the support. Therefore, fatigue test was carried out in this chapter with the maximum and minimum loads of 15 t and 3 t, respectively, and the fatigue loads were loaded at these high risk zones of drive axle housing. The fatigue testing device for the drive axle housing is shown in Fig. 12.2a. Before testing, the weld joint was polished to obtain a bright and smooth surface. When the number of fatigue cycles reached 100,000 and 300,000, the axle housing was placed on the support frame of the testing device. The surface MMM signals on testing lines 1 and 2 were detected offline, as shown in Fig. 12.2b. To avoid the error of manual testing, the stepper motor electronically

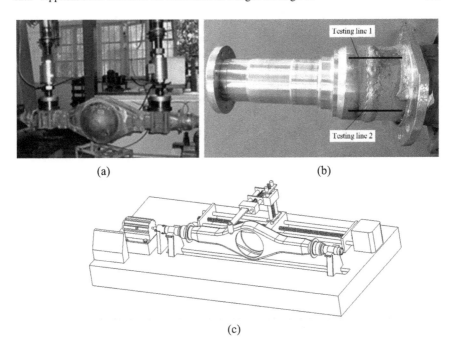

(a)   (b)

(c)

Fig. 12.2 Fatigue testing for the drive axle housing of a vehicle: **a** fatigue testing device; **b** testing line position; and **c** testing device

controlled balancing table was used to control the movement of the probe with a velocity of 8 mm/s, as shown in Fig. 12.2c.

The testing results are shown in Fig. 12.3. The overall variation amplitude of the normal component of the MMM signal $H_p(y)$ and its gradient value K both increase with the increase in the number of loading cycles. For the MMM signal of testing line 1, after 300,000 loading cycles, the MMM signal $H_p(y)$ shows obvious valley characteristics, and its gradient K reaches the maximum at a testing length of 40 mm, as shown in Fig. 12.3a, b. This is because the testing position of 40 mm is the weld joint of the drive axle housing where the high stress concentration and the serious damage may easily appear under the effect of cyclic loads. In addition, the number of zero-crossing points for $H_p(y)$ also increases with the increase in the number of loading cycles. There is only one zero-crossing point at the testing length of 40 mm after 100,000 cycles. However, when the loading cycles reach 300,000, the MMM signal $H_p(y)$ curve has four zero-crossing points. It indicates that the damage degree of drive axle housing obviously increases during the fatigue process. At the 300,000 loading cycles, the MMM signal gradient K curve fluctuates drastically at the testing length of 20-40 mm, which is consistent with the zero-crossing position of the MMM signal $H_p(y)$ curve.

Similarly, the amplitude of the normal component of the MMM signal $H_p(y)$ and its gradient value K at testing line 2 remain stable at the initial stage, as shown in

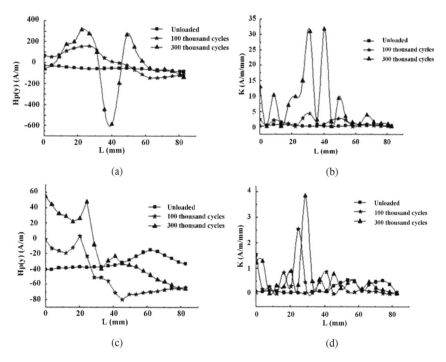

Fig. 12.3 Magnetic memory signals detected on the surface of the drive axle housing: **a** normal component $H_p(y)$ at testing line 1; **b** gradient value K at testing line 1; **c** normal component $H_p(y)$ at testing line 2; and **d** gradient value K at testing line 2

Fig. 12.3c, d. After 100,000 and 300,000 cycles, many peaks of the MMM gradient value K appear in the range of 20–40 mm, which is also consistent with the results of testing line 1. The most high risk zone of the drive axle housing is located at the testing length of 20–40 mm which is adjacent to the weld joint. Therefore, it is essential to pay more attention to this high risk zone to prevent accidents.

To verify the testing results by the MMM technique, X-ray testing was also used to measure the residual stress on the surface of the drive axle housing. Since we have known that the high risk zone is located at the weld joint of the drive axle housing, only the residual stress near the testing line of 20–40 mm is measured to reduce the testing cost. The residual stress distribution along testing line 1 is shown in Fig. 12.4. The residual stress fluctuates in a small range at the initial stage. Although the amplitude of the residual stress decreases with the increase in the number of loading cycles, the fluctuation degree of the residual stress increases with the increase in the number of loading cycles. Besides, the residual tensile stress gradually turns to the residual compressive stress during the fatigue process. And the overall residual compressive stress increases with the increase in the number of loading cycles, which is similar with the variation of MMM signals. It indicates that the MMM signal characteristic values have the ability to determine the stress concentration location and reflect the damage degree of the drive axle housing.

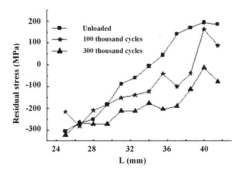

Fig. 12.4 Residual stress variation with the testing length of line 1

12.2.2 Relation Between MMM Signals and Deformation Degree

Under the effect of applied loads, not only can the stress concentration be induced, but also can the plastic deformation be formed. In order to evaluate the deformation of drive axle housing by MMM signals, six retired drive axle housings were randomly selected to measure their deformation based on the height difference. The axle housings were installed on the measuring frame, which could rotate along the axial direction at any angle. The half sleeve pipes at both ends of the axle housing were supported by a support ring. The center point of the reinforcing ring was measured by a height vernier caliper, as shown in Fig. 12.5. Then, the MMM signals were measured with the magnetic probe being fixed vertically in the vicinity of the weld joint along the busbar direction of the half-axle casing.

The deformation degree and MMM signal variation of retired drive axle housing #1~#6 were collected and shown in Table 12.1 and Fig. 12.6, respectively. The testing

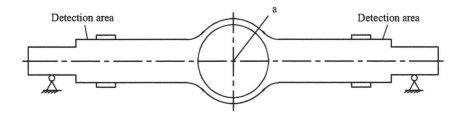

Fig. 12.5 Schematic diagram of drive axle housing deformation testing

Table 12.1 Deformation and peak value of the magnetic signal gradient K_{max} of retired axle housings

Retired axle housing	#1	#2	#3	#4	#5	#6
Deformation/mm	0.280	0.130	0.140	0.110	0.970	0.270
Gradient peak value K_{max} (A/m/mm)	0.80	0.57	0.55	0.46	10.50	0.63

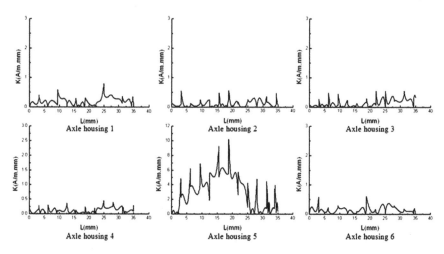

Fig. 12.6 Magnetic signal gradient value K of retired axle housings from #1 to #6

results show that the normal component of the MMM signal gradient K of sample #5 fluctuates sharply. The fluctuation degree of the MMM signal at the weld joint of the other five drive axle housings remains lower. Besides, the gradient peak value K_{max} of sample #5 is the largest compared with other samples and its peak is also located at the length range of 20–40 mm, which is consistent with the evaluation results in Fig. 12.3. Coincidentally, the deformation degree of sample #5 is also obviously larger than that of the other five axle housings. This means that the MMM signal characteristic value of the retired axle housing can also be used to reflect the plastic deformation degree under the same detection environment. The results show that the MMM is a feasible technique to quickly determine the high stress concentration and plastic deformation zones and qualitatively evaluate the damage degree of retired axle housing.

12.3 Application of MMM in the Evaluation of Fatigue Damage of Retired Hydraulic Cylinders

12.3.1 Threshold Determination Method for Remanufacturability Evaluation

Remanufacturability evaluation is the premise and basis of remanufacturing for retired components. However, lots of oil stain, rust, dust and other impurities usually adhere to the retired components, especially for the hydraulic cylinders, before cleaning or surface pretreatment. This makes it difficult to use the traditional nonde-structive testing to conduct the remanufacturability evaluation quickly and accurately

Table 12.2 The damage degree index K_{max} of retired hydraulic cylinders

Sample number	Average value (A/m/mm)	Standard deviation (A/m/mm)
#1	24.1	8.986381
#2	29.5	3.205897
#3	33.1	9.567015
#4	38.9	5.237684
#5	39.7	8.634087
#6	41.3	8.325434
#7	43.0	10.29563
#8	49.6	13.21783
#9	58.9	8.319322
#10	67.4	18.87338
Average	42.55	9.47

in engineering applications. Based on the previous researches of MMM technique, we propose a MMM threshold determination method to help solve the issue of remanufacturability evaluation in this section. First, ten retired hydraulic cylinders are selected randomly from a domestic construction machinery enterprise. The distributions of the normal component gradient value of the MMM signal K along the axial direction testing line of ten retired hydraulic cylinders are measured by MMM testing. Then, the maximum gradient value K_{max} on the testing line, which can be considered to be the damage degree index, is extracted to evaluate the remanufacturability based on statistical method. The MMM signals on each testing line are measured six times to calculate the average value and the standard deviation of the damage degree index K_{max}, as listed in Table 12.2.

It is assumed that the MMM signals follow a normal distribution. When the confidence level is 95%, the confidence interval is:

$$\left(\bar{X} \pm \frac{\sigma}{\sqrt{n}} z_{\alpha/2} \right) = \left(42.55 \pm \frac{9.47}{\sqrt{10}} \times 1.96 \right) \tag{12.1}$$

where $z_{\alpha/2}$ is the upper quantile of the standard normal distribution of the damage degree index. The final threshold value can be determined according to the required accuracy, as shown in Fig. 12.7. Based on the analysis results for the hydraulic cylinder, the expected confidence interval of the damage degree index is 36.68–48.42. Therefore, it can be believed that the measured damage degree index K_{max} for most retired hydraulic cylinders will be between 36.68 and 48.42 A/m/mm. Based on this, samples #8, #9 and #10 can be considered to be severely damaged, and they have no remanufacturability because their damage degree indexes K_{max} exceed the threshold value. Except for the maximum value of the normal component of the MMM signal gradient K_{max}, the damage degree can also be characterized by other indexes, such as the peak-to-peak value of the normal component of the MMM signal H_{p-p}, the peak-to-peak width of the normal component of the MMM signal

Fig. 12.7 Threshold
determination method

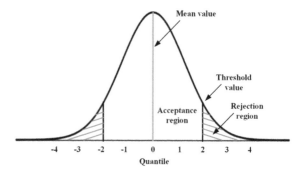

W_{p-p}, and the maximum value of the tangential component of the MMM signal $H_p(x)_{max}$. Furthermore, these damage degree indexes of the MMM signals can be affected by material properties, technological processes, detection positions, sizes and other factors. Therefore, it is necessary to modify the threshold value adaptively. The influencing factors and their effects are listed in Table 12.3 according to our experiments and practical experiences.

12.3.2 Experimental Verification

In this section, the retired hydraulic cylinders of steering cylinder 118E were selected and the remanufacturability was evaluated by MMM technique. Figure 12.8a and b show the appearance of hydraulic cylinder with and without obvious damage, respectively. The aforementioned MMM threshold determination method was used to analyze the remanufacturability of retired hydraulic cylinders. The normal and tangential components of the MMM signals of the hydraulic cylinders with obvious damage are collected, and one of the measurement results is shown in Fig. 12.9. The normal components of the MMM signals show the obvious peak-valley characteristic and the tangential components show the single peak at the damage zone. Based on the measured MMM signals, three different kinds of damage degree indexes are extracted and compared with the calculated threshold value, as shown in Table 12.4.

Similarly, the normal and tangential components of the MMM signals of the hydraulic cylinders without obvious damage are also collected, and one of the measurement results is shown in Fig. 12.10. Under the same range of vertical coordinate, it can be found that the magnetic signal curves in Fig. 12.10 are obvious smaller than those in Fig. 12.9, which indicates that the serious damages in hydraulic cylinder can induce the extensive spontaneous magnetic flux leakage on the parts' surface. The remanufacturability evaluation results are also shown in Table 12.5.

For the cylinder with obvious damage, all three measured damage degree indexes of the magnetic signal are larger than the calculated threshold value. Therefore, this retired hydraulic cylinder has lost its remanufacturing value. However, for the cylinder without obvious damage, all three measured damage degree indexes of the magnetic signal are smaller than the calculated threshold value. This indicates that this

Table 12.3 Threshold modified method and influencing factors for hydraulic cylinders

No.	Influencing factor	Content	Influence mechanism	Coefficient of threshold modification
1	Material properties	1. 45 steel (ASTM 1045) 2. 20 steel (ASTM 1020)	Various materials may have different carbon contents, so the magnetic domain movement also shows the difference, which leads to the variation in the macro magnetic signal	According to our experimental research, the amplitude of MMM signals in 20 steel are generally 1.67 times larger than that in 45 steel. Thus, we can take 45 steel as $K_{Material} = 1$, and 20 steel as $K_{Material} = 1.67$
2	Technological process	1. Chrome plating 2. Shot peening 3. Heat treatment	Chrome plating, shot peening and heat treatment can strengthen the materials, which can increase the ultimate strength by 33% and prolong their lives	The coefficient of threshold modification for processing technology can reach $K_{Process} = 1.3–2.0$.
3	Detection position	1. Axial surface 2. Circumferential surface 3. Step surface 4. Corner	Axial and circumferential surfaces are both regular surfaces, which can directly use the conclusions. However, the step surface will cause the fluctuation of magnetic memory signals, which needs to be processed after eliminating the rising trend. The corner can be measured by the tilt probe method	For the axial and circumferential surfaces, the coefficient is $K_{Surface} = 1$. For the other detection position, data processing should be conducted to eliminate the signal fluctuation
4	Size	1. The detection length is much smaller than the component size 2. The test length has the same order of magnitude as the component size	When the detection length is close to the component size, the magnetic signal will increase rapidly due to the size effect	The distortion signals at the both ends of the component are removed during data processing

(a)   (b)

Fig. 12.8 Retired hydraulic cylinder: **a** with obvious damage and **b** without obvious damage

Fig. 12.9 Variation in the
magnetic signals of
hydraulic cylinders with
obvious damage

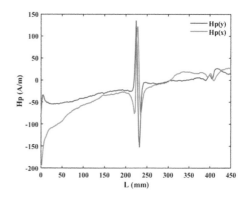

Table 12.4 Evaluation results for hydraulic cylinders with obvious damage

No.	Index	Threshold	Measured value	Evaluation results
1	Maximum normal component gradient K_{max} (A/m/mm)	48.42	54	Remanufacturability Negative
2	Peak-to-peak value of the normal component $H_{p\text{-}p}$ (A/m)	79.27	285	
3	Tangential component maximum $H_p(x)_{max}$ (A/m)	36.88	123	

retired hydraulic cylinder still has a larger remanufacturing value and can be repaired. The evaluation results indicate that the proposed damage degree index and the MMM threshold determination method can effectively evaluate the remanufacturability of retired hydraulic cylinders.

Fig. 12.10 Variation in the magnetic signals of hydraulic cylinders without obvious damage

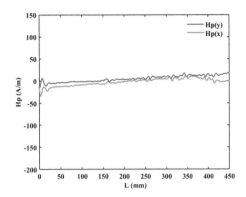

Table 12.5 Evaluation results for hydraulic cylinders without obvious damage

No.	Index	Threshold	Measured value	Evaluation results
1	Maximum normal component gradient K_{max} (A/m/mm)	48.42	6	Remanufacturability Positive
2	Peak-to-peak value of the normal component H_{p-p} (A/m)	79.27	46	
3	Tangential component maximum $H_p(x)_{max}$ (A/m)	36.88	29	

12.4 Conclusions

The MMM testing method can quickly determine the high risk zone of the drive axle housing and the stress concentration based on the magnetic signal characteristic value. The normal component of the magnetic signal curve $H_p(y)$ at the damage zone has abrupt change, and its gradient K has an abnormal peak value. The peak value of the magnetic signal gradient K_{max} corresponds to the deformation degree of the retired axle housing, and the maximum gradient value K_{max} increases with increasing deformation. In addition, the magnetic signal characteristic value of the drive axle housing also increases with the increase in the number of loading cycles. Therefore, the service life of the drive axle housing can be expressed by the characteristic value K_{max} or the deformation degree.

The maximum gradient value K_{max} of the normal component of the magnetic signal is extracted as the damage degree index, which can be used as the basis for the remanufacturability evaluation for the retired hydraulic cylinder. Because the magnetic signals have a trend of stable variation and the characteristics of signal superposition, the average value and the standard deviation of the damage degree index were calculated based on the statistical method. Then, the MMM threshold determination method for remanufacturability evaluation was established and modified. Finally, the retired hydraulic cylinder with and without obvious damage is taken as an example to verify the feasibility of the method.

However, the MMM signal is very weak and can be easily affected by various environments. Furthermore, the structure, working conditions, residual stress, deformation and environmental magnetic field of the retired drive axle housing and hydraulic cylinders are also complex, and it is currently difficult to formulate a unified standard to quantify the damage degree. To improve the evaluation accuracy, it is necessary to combine MMM testing with other nondestructive testing technologies, such as eddy current, Barkhausen noise, and X-ray methods, to comprehensively evaluate the remanufacturability of retired components including axle housing and cylinders in the future.

References

1. C.C. Li, L.H. Dong, H.D. Wang et al., Metal magnetic memory technique used to predict the fatigue crack propagation behavior of 0.45%C steel. J. Magn. Magn. Mater. **405**, 150–157 (2015)
2. Z.C. Qian, H.H. Huang, Coupling fatigue cohesive zone and magnetomechanical model for crack detection in coating interface. NDT & E Int. **105**, 25–34 (2019)
3. C. Ni, L. Hua, X.K. Wang, Crack propagation analysis and fatigue life prediction for structural alloy steel based on metal magnetic memory testing. J. Magn. Magn. Mater. **462**, 144–152 (2018)

Chapter 13
Evaluation of the Repair Quality of Remanufactured Crankshafts

13.1 Introduction

Due to the inconformity of the failure mode or damage degree in the remanufacturing cores as shown in Chap. 12, the corresponding repair process parameters in remanufacturing may also be different. Therefore, the mechanical properties of remanufactured products are affected by the status of cores, which can lead to quality fluctuations after repair. It is necessary not only to evaluate the damage degree of the cores before remanufacturing, but also to evaluate the repair quality after remanufacturing. Only in this way can unqualified remanufactured products be prevented from reaching the market.

In this chapter, we take remanufactured crankshaft, which is one of the key components of an engine, as an example to show the metal magnetic memory (MMM) testing method in the evaluation of repair quality after remanufacturing. During long-term operation, each shaft neck will slide at a high speed under a high specific pressure, which can easily lead to surface wear. It is not wise to scrap the worn crankshaft directly because of its high added value, which accounts for 10–20% of the cost of the whole machine. Therefore, we often use advanced surface engineering technology to repair worn crankshafts. After remanufacturing, the service life of the crankshaft can be prolonged, and the performance can be improved to some extent.

Among the common remanufacturing technologies, plasma transferred arc welding (PTAW) has been widely used in the repair and strengthening of mechanical parts due to its advantages of high productivity, low manufacturing cost, and good forming quality [1, 2]. However, during PTAW, the performance of the remanufactured coating is controlled by many process parameters, such as the cladding current, the scanning speed, and the powder flow rate. Due to the combined effects of plastic deformation, temperature gradients and metallurgical changes, the residual stress of cladding coating and heat-affected zone is complex. Furthermore, the residual stress will greatly reduce the fatigue life of ferromagnetic parts, induce crack initiation, and even cause serious accidents. Therefore, to obtain a wear-resistant remanufactured coating with excellent performance and low residual stress, it is necessary to

© Science Press 2021
H. Huang et al., *Metal Magnetic Memory Technique and Its Applications in Remanufacturing*, https://doi.org/10.1007/978-981-16-1590-0_13

optimize the PTAW process parameters first. Besides, quickly and accurately evaluating the residual stress of the coating is also the key to ensuring the quality of the remanufactured parts.

Based on the contents of Chaps. 9–11 in this book, the thermal residual stress, the coating performance and the interface cracks can be well evaluated by MMM testing. Therefore, the abovementioned research results and conclusions can also provide references in the evaluation of remanufactured crankshafts. In this chapter, the orthogonal test was first carried out on PTAW in Sect. 13.2. Then, the process parameters were optimized, and the cladding coating with the best wear resistance was analyzed in Sect. 13.3. Finally, a retired engine crankshaft was repaired, and its surface MMM signals were measured and studied to reflect the residual stress distribution in the cladding coating and the heat-affected zone in Sect. 13.4. In summary, this research can provide guidance for the quality evaluation of remanufactured crankshafts.

13.2   Repair Process in Remanufacturing

The substrate material for PTAW was 45 steel. The substrate was cut into 100 mm × 100 mm × 10 mm pieces and then ground and cleaned with acetone. Ni60 alloy powder was selected and dried at 120 °C for 2 h. The chemical compositions of the 45 steel substrate and Ni60 powder are given in Table 5.1 and Table 9.1, respectively.

The nickel-based coatings were deposited on the substrate by a PTA-400E4-ST PTAW equipment, which was designed by the Wuhan Research Institute of Materials Protection, as shown in Fig. 13.1. Through preliminary MMM detection for the

(a)   (b)

Fig. 13.1 Overview of the experiment: **a** PTAW equipment; and **b** experimental process

Table 13.1 Control factors with their levels for orthogonal L9 experimental design

Control factors	Unit	Levels		
		1	2	3
Cladding current	A	100	110	120
Scanning speed	mm/min	50	60	70
Powder flow rate	g/min	14	18	22

cladding coating, the appropriate range of the process parameters was determined by excluding the bad parameters, which can lead to obvious cracks and large residual stress. Then, the orthogonal experiment was designed to help understand the effect of the process parameters on the cladding quality. Table 13.1 presents the selected control factors [3] and their levels for the experiments. The other process parameters were fixed in this experiment with a plasma voltage of 28 V, torch-to-surface distance of 10 mm, swing width of 14 mm and swing speed of 1400 mm/min. Ar was taken as a plasma gas, carrier gas and shielding gas with flow rates of 300 L/h, 300 L/h and 800 L/h, respectively.

After PTAW, the samples were cut via wire electrical discharge machining. The cross-sections were etched by dilute aqua regia (50 ml HCl + 25 mL HNO$_3$ + 25 mL H$_2$O) to reveal the microstructure of the coating. The microstructure of the coating was analyzed by optical microscopy (OM) (OPTIKA XDS-3MET). The Vickers microhardness (HV) was determined with an HVS-1000A microhardness tester by applying a load of 1000 g for 15 s to produce a series of indentation marks separated by 300 μm. Dry sliding wear tests were carried out on a self-developed pin-on-block MM-200 wear tester under a load of 135 N and a sliding speed of 25 mm/s for 2 h at 25 °C. The size of each pin was cut into 14 mm (height) × 6 mm × 6 mm pieces. Before the wear test, the surface of the samples was ground to Ra 0.8. Dantsin Trimos TR-SCAN 3D surface topography device was used to measure the volume loss of the samples.

13.3 Evaluation of the Repair Quality of the Remanufactured Coating

13.3.1 Optimization of the Processing Parameters

To investigate the effect of the process parameters on the coating wear resistance, experimental trials were conducted by orthogonal tests [4], and the data were analyzed based on variance analysis. Table 13.2 lists the various combinations of the process parameters and their effects on the properties of the nickel-based coating. Then, the variance analyses of the microhardness and volume loss were conducted and the results are shown in Table 13.3 and Table 13.4, respectively, based on the

Table 13.2 Experimental trials based on orthogonal design and their effects

No.	Cladding current (A)	Scanning speed (mm/min)	Powder flow rate (g/min)	Coating properties	
				Microhardness $(HV_{1.0})$	Volume loss $(10^{-3}\ mm^3)$
1	100	50	14	445.4	10.23
2	100	60	18	475.5	4.18
3	100	70	22	562.1	5.18
4	110	50	18	485.4	4.91
5	110	60	22	507.7	6.30
6	110	70	14	393.0	9.38
7	120	50	22	533.2	6.11
8	120	60	14	382.9	11.85
9	120	70	18	560.3	6.35

Table 13.3 Variance analysis of the microhardness

Sources of variance	Sum of the squares of deviations	Degrees of freedom	Variance	F-measure	F_a-measure	Significance
Cladding current	1954.1	2	977.05	0.93	6.94	No
Scanning speed	3835.2	2	1917.6	1.83	6.94	No
Powder flow rate	26925.1	2	13462.55	12.83	6.94	Yes
Error	2243.3	2	1121.65			
Error correction	4197.4	4	1049.35			

Table 13.4 Variance analysis of the wear volume loss

Sources of variance	Sum of the squares of deviations	Degrees of freedom	Variance	F-measure	F_a-measure	Significance
Cladding current	4.12	2	2.06	3.75	6.94	No
Scanning speed	0.37	2	0.19	0.35	6.94	No
Powder flow rate	50.40	2	25.20	45.82	6.94	Yes
Error	1.82	2	0.91			
Error correction	2.19	4	0.55			

data in Table 13.2. It can be seen from Tables 13.3 and 13.4 that the powder flow rate has a substantial influence on the microhardness and volume loss, while the cladding current and scanning speed have no substantial influence on the properties of the coating.

13.3.2 Effect of the Processing Parameters on the Microstructure

Figure 13.2 shows the microstructures of the coating with different powder flow rates at a cladding current of 100 A and scanning speed of 70 mm/min. The microstructures of the coating vary substantially with different powder flow rates. Figure 13.2a, b show the microstructures of the near-fusion zone and the near-surface zone of the coating under the 14 g/min powder flow rate. The bottom fusion line is relatively straight, and there are a large number of columnar crystals growing perpendicular to the fusion line in the near-fusion zone. A large dendritic skeleton structure appears near the surface area [5], and the main growth direction is opposite to the direction of heat flow. Figure 13.2c, d show the microstructures of the near-fusion zone and near-surface zone of the coating under the 18 g/min powder flow rate. The bottom fusion line is also relatively straight, and there are a large number of flocculent and fine dendrites in the near-fusion zone. With increasing distance from the fusion line, dendrite growth tends to be stable and many long strips appear. Figure 13.2e, f show the microstructures of the near-fusion zone and near-surface zone of the coating under the 22 g/min powder flow rate. The profile of the bottom fusion line is not clear, and the near-fusion zone grows an equiaxed crystal structure with uniform distribution. With increasing distance from the fusion line, the dendritic crystal structure disappears and changes to equiaxed crystal gradually.

In Fig. 13.2, it can be seen that with increasing powder flow rate, the dendrite spacing decreases, and the microstructure of the coating becomes fine. The coating microstructure changes gradually from columnar crystals to dendrites and equiaxed crystals. Under the same cladding current and scanning speed, the energy input to the molten pool per unit time remains unchanged, but the change in powder flow rate will lead to different energies obtained per unit mass of powder. From the overall trend, the larger the energy per unit mass of powder is, the more fully the alloy powder dissolves, which leads to a larger melting amount of the substrate or a larger dilution rate. With an increase in the powder flow rate, the energy obtained per unit mass of powder decreases, while the temperature gradient and the solidification speed increase, which leads to an insufficient quantity of columnar crystals. However, when the heat input per unit mass of powder decreases, the melting of alloy powder is incomplete. Nucleation appears in the subsequent solidification process of the molten pool, which further produces fine grains in the microstructure. Therefore, the columnar crystal structure decreases and the dendrite and equiaxed crystal structure

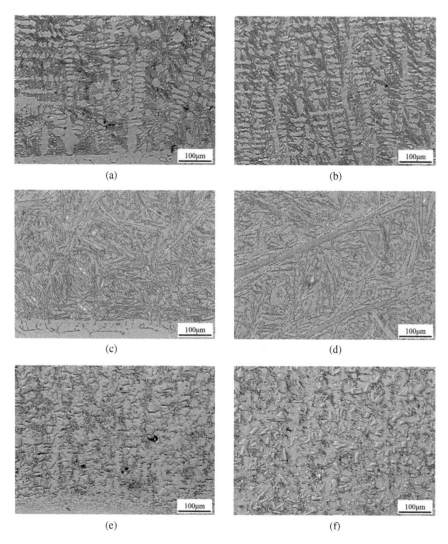

Fig. 13.2 Microstructure of the coatings with different powder flow rates: **a** near-fusion zone, 14 g/min; **b** near-surface zone, 14 g/min; **c** near-fusion zone, 18 g/min; **d** near-surface zone, 18 g/min; **e** near-fusion zone, 22 g/min; and **f** near-surface zone, 22 g/min

increases with increasing powder flow rate in a certain range. In this situation, the performance of the cladding coating is greatly improved [5].

In summary, when the powder flow rates were 18 g/min and 22 g/min, the microstructures of the coatings were dendrites and equiaxed crystals, respectively. This means that the mechanical properties of the coating were good under this process parameter. However, when the powder flow rate was 14 g/min, the microstructure of the coating was columnar crystals, and the mechanical properties were poor.

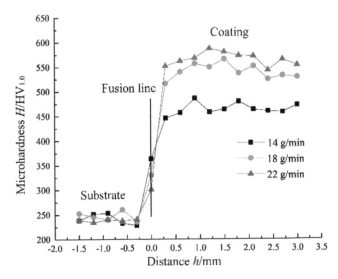

Fig. 13.3 Microhardness of coatings with different powder flow rates

13.3.3 Effect of the Processing Parameters on the Microhardness

The microhardness distribution along the cross-sections of nickel-based coatings with three different powder flow rates is presented in Fig. 13.3. The value of the microhardness in the substrate 45 steel is 240 HV, while the value is between 460 HV and 560 HV in the cladding zone, which is attributed to the generation of finer dendrites and a large number of intermetallic compounds based on the above microstructure analysis [6]. The microhardness in the coating with a powder flow rate of 22 g/min is approximately 560 HV, while the values are approximately 540 HV and 460 HV for powder flow rates of 18 g/min and 14 g/min, respectively. This indicates that the different dendritic structures of the coating under various powder flow rates show different mechanical properties. Within a certain range, the microstructure of the coating changes from columnar to dendritic and equiaxed grains with increasing powder flow rate. In addition, as the grains become finer, the heat damage degree of the strengthening phase decreases, and the effect of the solid solution and fine grain strengthening increases, so the microhardness of the coating increases substantially.

13.3.4 Effect of the Processing Parameters on the Wear Resistance

Figure 13.4 shows the morphology of the wear surface of the substrate and nicked-based coatings with different powder flow rates. It can be seen that both the substrate

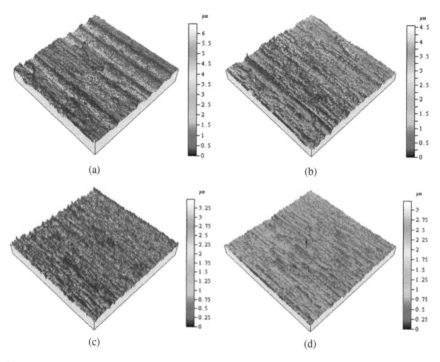

Fig. 13.4 Worn morphology of the substrate and coatings with different powder flow rates: **a** substrate; **b** 14 g/min; **c** 18 g/min; and **d** 22 g/min

and the coatings show the characteristics of abrasive wear after the wear test. There are deep and wide furrows on the substrate surface, and the furrow spacing is not uniform. The furrows on the worn surface of the coating are shallow, compact and evenly distributed compared with the substrate, which indicates that the coating has better wear resistance than the substrate. Moreover, with increasing powder flow rate, the abrasive wear degree decreases, and the plough groove on the worn surface also becomes shallow and narrow. This indicates that the wear resistance of the coating increases, which is consistent with the change trend of the microhardness mentioned above. When the powder flow rate is 22 g/min, the coating has the best wear resistance.

The volume loss during dry friction under different powder feeding rates is shown in Fig. 13.5. The volume losses of the substrate and the coating with powder flow rates of 14, 18 and 22 g/min are 17.22×10^{-3}, 8.47×10^{-3}, 5.55×10^{-3} and 5.02×10^{-3} mm^3, respectively. The volume loss decreases with increasing powder flow rate. It can be also found that the volume loss of the coating exhibits little difference for the powder flow rates of 18 and 22 g/min.

Figure 13.6 shows the friction coefficient curves of the substrate and three types of nicked-based coatings. It can be noted that the average friction coefficient value

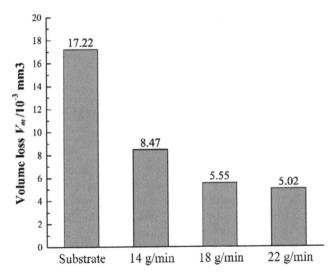

Fig. 13.5 Volume loss of the substrate and coatings with different powder flow rates

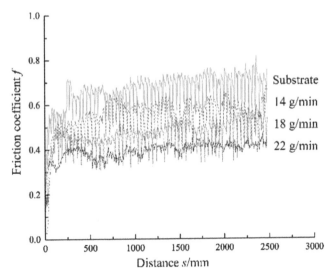

Fig. 13.6 Friction coefficient of the substrate and coatings with different powder feed rates

of the substrate is 0.66, while the values of the coatings with powder flow rates of 14, 18 and 22 g/min are 0.55, 0.47 and 0.41, respectively. It is known that the friction coefficient is inversely proportional to the corresponding microhardness. Therefore, the coatings with a powder flow rate of 22 g/min have excellent wear resistance, and their change in friction coefficient is the most stable, which indicates that the performance of the coatings are substantially improved.

In summary, when the powder flow rate is 22 g/min, the worn surface of the coating has a shallow and narrow furrow, and the wear volume loss and the friction coefficient are the smallest, which means the coating has great wear resistance.

13.4   Repair Quality Evaluation Based on MMM Measurement

After a period of time in service, the spindle neck of the engine crankshaft of a certain enterprise has a groove with a wear depth of approximately 0.5 mm. The surface damage of the crankshaft is repaired by PTAW with a welding current of 100 A, welding speed of 70 mm/min and powder feeding flow rate of 22 g/min based on the optimal process parameters mentioned above. Before remanufacturing, ultrasonic cleaning was carried out on the crankshaft to remove sundries and stains on the surface. To reduce the impact of the thermal effect of cladding on the crankshaft dimension accuracy, the sample is preheated to 200 °C before repair, and then PTAW is carried out on the surface wear parts of the spindle neck.

To evaluate the repair quality, a homemade MMM detection instrument is used to detect the state of the crankshaft pre- and post-remanufacturing. Based on the characteristic value of the magnetic signals, MMM testing technology can determine the residual stress or other defects of mechanical parts. A schematic diagram of MMM detection is shown in Fig. 13.7. The lift-off height of the detection probe is set to 1 mm, the detection speed is kept at 5 mm/s, and the MMM signals along the circumference of the spindle neck are measured by rotating the crankshaft.

The tangential component $H_p(x)$ and the normal component $H_p(y)$ of the MMM signals before and after remanufacturing are obtained as shown in Fig. 13.8. There are substantial differences in the MMM signals between the pre- and post-repair periods of the crankshaft. In Fig. 13.8a, the tangential components $H_p(x)$ of the MMM signals fluctuate overall at approximately 0 A/m before repair, while there is a distinct peak at an angle of 120°, where the peak value is 95 A/m. After repair, the curve only slightly fluctuates at approximately 0 A/m, and no obvious peaks appear. Similarly, in Fig. 13.8b, the normal components $H_p(y)$ of the MMM signals change positively and negatively at an angle of 120° before repair, and the peak and valley values are 351 A/m and −376 A/m, respectively. There is also no obvious signal fluctuation after repair. Based on the results in Chap. 5, the distortion of MMM signals of the worn spindle neck of the engine crankshaft before repair is induced by the tribo-magnetization. And the damage is mainly located at the angle of 120° along the circumference. After remanufacturing, the stable MMM signals show that the defects in the crankshaft are repaired and the heat residual stress is small and distributed uniformly. The evaluation results of MMM technique indicate that the repair quality of the crankshaft under the optimal process parameters meets the standards of remanufactured products.

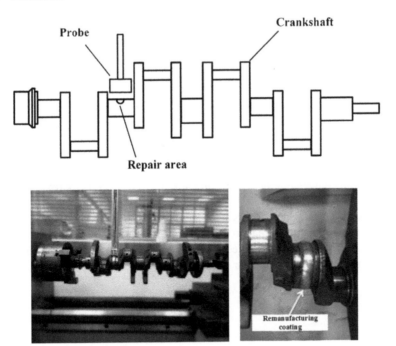

Fig. 13.7 MMM detection for the remanufacturing crankshaft neck

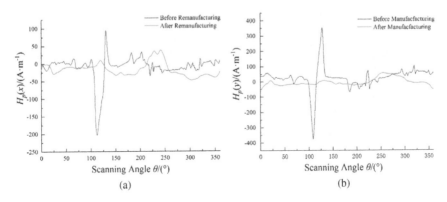

Fig. 13.8 Magnetic field strength before and after remanufacturing: **a** tangential component $H_p(x)$ and **b** normal component $H_p(y)$

13.5 Conclusions

A high hardness and wear resistance cladding coating was prepared by PTAW technology which can be successfully applied to the remanufacturing of engine crankshafts. The orthogonal test was used to optimize the PTAW process parameters.

The effects of the process parameters such as the cladding current, scanning speed and powder feeding flow rate on the microhardness and volume loss were studied. It was found that the powder feeding flow rate is the most substantial factor affecting the microhardness and volume loss. Within a certain range, the microstructure of the cladding layer changes from columnar crystals to dendritic and equiaxed crystals with increasing powder feeding flow. After repair, the mechanical performances, such as the hardness and wear resistance of the coatings, are greatly improved. The optimal process parameters are a spray welding current of 100 A, a spray welding speed of 70 mm/min and a powder feeding flow rate of 22 g/min. The cladding coating prepared under the optimal parameters has a good metallurgical bond with the equiaxed crystal microstructure. The average microhardness is approximately 560 $HV_{1.0}$, the average friction factor is 0.41, and the wear volume is 5.02×10^{-3} mm^3, which substantially improves the wear resistance. Finally, the worn engine crankshaft neck is repaired with the optimal process parameters. The results of MMM detection show that there are no obvious defects or high residual stress in the crankshaft after remanufacturing and that the repair quality can meet the product standards.

References

1. Y.L. Ge, D.W. Deng, X. Tian et al., Influence of parameters on microstructure and microhardness of Ni60 alloy hardfacing by plasma transferred arc welding. China Surf. Eng. **24**(5), 26–31 (2011)
2. D.W. Deng, R. Chen, H.C. Zhang, Present status and development tendency of plasma transferred arc welding. J. Mech. Eng. **49**(7), 106–112 (2013)
3. S. Li, J.J. Hu, G.Q. Chen et al., Microstructure and friction and wear properties of plasma surfacing layer of Ni-based alloy. Mater. Mech. Eng. **37**(6), 72–77 (2013)
4. O.N. Çelik, Microstructure and wear properties of WC particle reinforced composite coating on Ti6Al4V alloy produced by the plasma transferred arc method. Appl. Surf. Sci. **274**, 334–340 (2013)
5. M. Ulutan, K. Kiliçay, O.N. Çelik et al., Microstructure and wear behaviour of plasma transferred arc (PTA)-deposited FeCrC composite coatings on AISI 5115 steel. J. Mater. Process. Technol. **236**, 26–34 (2016)
6. F. Fernandes, B. Lopes, A. Cavaleiro et al., Effect of arc current on microstructure and wear characteristics of a Ni-based coating deposited by PTA on gray cast iron. Surf. Coat. Technol. **205**(16), 4094–4106 (2011)

Chapter 14
Development of a High-Precision 3D MMM Signal Testing Instrument

14.1 Introduction

In recent years, metal magnetic memory (MMM) detection technology has been applied to many engineering fields. Due to the expansion of market demand, a large number of MMM testing instruments have been developed by domestic and foreign researchers. Different types of MMM testing instruments have also been invented for different application situations, such as the petrochemical industry, railways, and aerospace.

At present, there are mainly two large companies that produce and sell commercial MMM testing instruments. The international one is the Energodiagnostika Co. Ltd. of Russia, whose TSC series products can often be used in the diagnosis of boilers, petroleum pipelines, bridge construction and construction machinery. This type of instrument can obtain spontaneous magnetic flux leakage information on the surface of the ferromagnetic component, which can help to determine the defect or the stress concentration [1]. The domestic one in China is Xiamen Edson Electronics Co., Ltd., which develops the EMS series MMM intelligent testing instrument. It can not only effectively detect MMM signals but also combine other nondestructive testing (NDT) technologies, such as eddy current and magnetic flux leakage, to improve the evaluation reliability. Its products have been successfully applied in the damage detection of aircraft blades and landing gear. The abovementioned two instruments have been used by many scientific research institutes. For example, in 1999, the researchers of the Northeast Electric Power Research Institute used the TSC-1M-4 type product for MMM testing of power plant boiler pipes. Then, the research team from Harbin Institute of Technology also used this product to study the MMM signals of ferromagnetic materials during the fatigue process. The research teams from Hefei University of Technology and Yanshan University analyzed the variation in MMM signals with the EMS-2003 MMM testing instrument and successfully evaluated the weld defects.

In addition, there are also many other institutes that are devoted to independent development of MMM testing instruments. Germany Fraunhofer IZFP proposed a

© Science Press 2021
H. Huang et al., *Metal Magnetic Memory Technique and Its Applications in Remanufacturing*, https://doi.org/10.1007/978-981-16-1590-0_14

composite inspection method and developed a new MMM testing instrument. By combining this testing instrument with an intelligent algorithm, it is possible to effectively evaluate the residual stress and cracks of mechanical components [2]. Zhang et al. designed an array of MMM sensors with Hall elements and built a simplified MMM testing instrument that can detect the crack growth of components with complex shapes or large surface areas. The experimental results show that this type of MMM testing instrument can effectively monitor crack growth [3]. The research team of Tsinghua University developed a handheld MMM testing instrument within high-sensitivity magnetoresistive sensors, and it can detect the weak magnetic signals of defects from train rails and pipelines [4]. Ren et al. developed a new MMM testing instrument capable of two-dimensional detection by combining a microprocessor with a tunnel magnetoresistance (TMR) sensor. The performance of the instrument has been verified, and a quantitative analysis of the stress concentration was performed through the Lissajous diagram [5].

However, the types of MMM testing instruments are still few, and their functions are relatively simple at present. Therefore, the disadvantages of insufficient signal collection, low detection accuracy and heavy instrumentation limit the application range of MMM detection in remanufacturing.

To realize the accurate evaluation of early damage in remanufacturing cores or repair quality in the remanufacturing process, a new three-dimensional MMM testing instrument with an embedded weak magnetic detection system based on an anisotropic magnetoresistance (ARM) sensor is developed in this chapter. The development purpose of this instrument is to quickly and accurately measure the magnetic field information on the surface of the ferromagnetic component and realize the real-time waveform display of the three-axis data of the magnetic field at the damage zone. In addition, it can help to promote the application of the instrument in remanufacturing. In this chapter, the system composition and the data processing flow of this testing instrument are introduced in Sect. 14.2. The hardware composition and the software design are described in Sect. 14.3. In Sects. 14.4 and 14.5, we calibrate and use the self-developed testing instrument to collect, display and analyze the MMM signals in damaged ferromagnetic materials. Finally, the performance of the self-developed instrument is compared with two other common commercial MMM testing instruments in Sect. 14.6.

14.2 Framework of the Detection System

The diagram of the embedded MMM detection system based on the AMR sensor is shown in Fig. 14.1. The MMM testing instrument contains a variety of circuit modules, such as the HMC5883L sensor module, embedded core control module, power module and liquid crystal display (LCD) module. The HMC5883L module includes a three-dimensional magnetic-sensitive bridge circuit and a signal processing circuit. The embedded core control module is composed of an ARM core board and a data processing module that mainly consists of magnetic signal acquisition and defect inversion. The MMM signal is converted into a corresponding

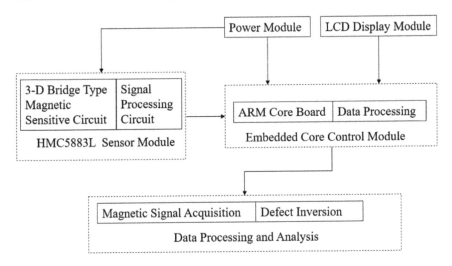

Fig. 14.1 Diagram of the MMM detection system

digital signal by the sensor module through the probe, and it is transmitted to the embedded core control module via the IIC bus, which can carry out corresponding signal processing. The MMM signal is finally displayed on the LCD screen.

14.3 Detailed Processes of Instrument Development

14.3.1 Hardware Design

• Probe

The circuit schematic diagram of the MMM detection probe is shown in Fig. 14.2a. The AMR sensor module HMC5883L has high sensitivity and good linearity, and its resolution reaches 0.4 A/m. The HMC5883L module adopts the XC6206 voltage stabilizing module to supply power and communicates with the external module through the IIC bus to realize data transmission. XC6206 is a type of high precision low dropout voltage regulator fabricated in the complementary metal oxide semiconductor (CMOS) technology. Its output range is adjustable within the range of 1.2–5.0 V, and the adjustment accuracy is 0.1. The chip is compatible with ceramic capacitors, which can save space for the baseboard and cost for the peripheral components.

The module wiring port mainly includes the power line, the IIC bus data line and the control line. The probe accessory circuit is integrated in the USB line. The USB shield line can decrease the effect of the clutter signal. The digital interface circuit is used to connect USB shield line to the host. The chip casing is made of plexiglass, which should be assembled under the premise of ensuring the accuracy of the sensor.

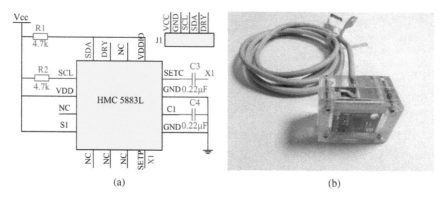

Fig. 14.2 Probe design: **a** the circuit schematic; and **b** the packaged probe

The sensor and its peripheral auxiliary circuit are installed in the designed insulating casing, as shown in Fig. 14.2b.

- Detection system core board hardware

The OKMX6UL developed by Forlinx Embedded is used as the hardware development platform of the MMM detection system. The development board adopts the microprocessor designed by Freescale based on ARM's authorized Cortex-A7 architecture, which has the advantages of excellent performance, low power and low cost. The processor can reach a maximum frequency of 528 MHz, and the power supply voltage is 4.5–5.5 V. This core board includes the RAM (512 MB LvDDR3), ROM (4 GB eMMC), and rich interface resources such as USB-OTG, LCD, a digital camera, JTAG simulation debugging, IIC bus, debugging serial, a network interface, a USB, an SD card, etc. The debugging serial interface is used to connect the system to the debugging software Secure CRT. The weak magnetic information is collected by the IIC bus interface. The LCD interface can help to realize the input of the parameters and the display of MMM signals. Multimachine communication is completed by the Ethernet interface and programming.

14.3.2 Software Design

- Qt application development

The excellent cross-platform features of Qt allow it to be deployed on different operating systems. Furthermore, Qt/Embedded provides cross-platform development tools that make embedded development fast and easy. Two versions of the embedded development environment are needed for compilation. The first one is to simulate the executable program by writing Qt application program on the X86 version of virtual machine. The second one is to run the executable program at embedded system by writing Qt application program on the ARM version. The Qt compiler should be added to the corresponding integrated development environment after compilation.

- Data acquisition

The embedded device driver provides the access mechanism and the device interface. The IIC bus is used to control the configuration of the registers by manipulating the serial clock (SCL) and serial data (SDA) in the HMC5883L. The magnetic signal data is obtained by accessing the device node of the embedded system. After data acquisition, the dimensional transformation of data is required.

- Interface display program

The display module is designed by combining digital images with waveform images. The former shows the size of the data, and the latter plots a real-time curve of the collected magnetic field data. The corresponding relationship between the time and the collected magnetic field data is displayed, which reflects the changing trend of the magnetic signals.

The programming flow chart is shown in Fig. 14.3. After the device is powered on, the interface is initialized at first, and three arrays are created to record the values of the MMM signals from the three directions, X, Y, and Z, of the sensors. In addition, a graphical display interface is also created at the same time. After pressing the acquisition button, the data acquisition program is executed, and finally, the real-time display of the MMM signal is realized. In addition, the entire display interface functions also include saving and screen capture.

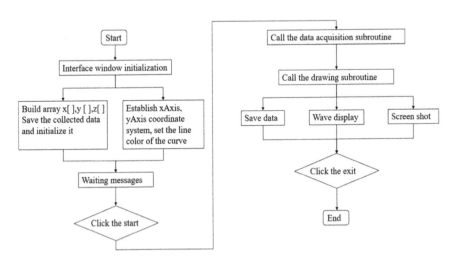

Fig. 14.3 Flow chart of the program

A simple data acquisition graphical interface is designed by writing the main function of the Qt application program based on the Qt Creator integrated development environment. An executable file is formed after kits-mx6 are embedded into Qt cross-compiling. Finally, it can be run by the network file system (NFS) mounted on the target board through the built local area network (LAN) system.

14.4 Calibration of Self-developed Instrument

14.4.1 Static Performance of the Instrument

A static magnetic field is generated based on high-precision Helmholtz coils, whose schematic diagram is shown in Fig. 14.4. The measurable value of the self-developed instrument is calibrated by calculating the theoretical constant magnetic field at the center of coils. The magnetic field on the central axis of a single coil is expressed according to the Biot-Savart law:

$$B = NI\frac{\mu_0 R^2}{2(R^2 + x^2)^{3/2}} \tag{14.1}$$

where N is the number of coil turns, x is the distance between the space point and coil center, μ_0 is the vacuum permeability, R is the average radius of the Helmholtz coils, and I is the electric current of the coils. Based on the superposition principle, the magnetic field on the central axis of the Helmholtz coils is:

$$B = \frac{1}{2}\mu_0 NIR_0\left\{\left[R^2 + \left(x + \frac{R}{2}\right)\right]^{-\frac{3}{2}} + \left[R^2 + \left(-x + \frac{R}{2}\right)\right]^{-\frac{3}{2}}\right\} \tag{14.2}$$

Fig. 14.4 Schematic diagram of the Helmholtz coils

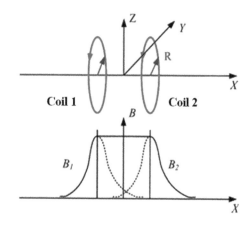

Table 14.1 Calibrated performance parameters of the MMM testing instrument

Performance parameters	Calculation equation	Tangential component	Horizontal component	Normal component
Linearity	$\delta_L = \pm \frac{\Delta Y_{max}}{Y_{F.S}} \times 100\%$	2.364%	2.513%	1.93%
Sensitivity	$k = \pm \frac{\Delta y}{\Delta x}$	2.667 LSB/A/m	2.733 LSB/A/m	2.614 LSB/A/m
Resolution	$\begin{cases} \delta_f = \frac{\Delta}{k} = B_2 - B_1 \\ \Delta = y_2 - y_1 \\ y = kB + b \end{cases}$	0.36 A/m	0.35 A/m	0.37 A/m
Repeatability	$\delta_H = \frac{\Delta H_{max}}{Y_{F.S}} \times 100\%$	1.15%	0.78%	0.75%
Hysteresis	$\delta_m = \frac{\Delta m_{max}}{Y_{F.S}} \times 100\%$	0.97%	0.66%	0.60%
Overall accuracy	$\delta = \sqrt{\delta_L^2 + \delta_H^2 + \delta_m^2}$	2.80%	2.71%	2.16%

The calibration system of the MMM testing instrument is established, including the Helmholtz coils, a Rigol DP 832 linear DC power generator, a gauss meter, and a magnetic field acquisition system. By adjusting the position and direction of the Helmholtz coils, the generated magnetic field is perpendicular to the geomagnetic field. Different theoretical magnetic field amplitudes can be adjusted by changing the current of the power supply. Three different components of the magnetic field in space are collected for calibration compared with the theoretical value, and the calibration results are listed in Table 14.1. The key performance parameters, such as the sensitivity, resolution and accuracy, all meet the requirements.

14.4.2 Ability to React to the Geomagnetic Field

It is very important to determine the ability of self-developed instrument to react to the geomagnetic field because the MMM signals are measured under the effect of the geomagnetic field. Based on the theoretical calculation of the world magnetic model, the geomagnetic field parameters at our laboratory position are listed in Table 14.2. The geomagnetic field can be described by seven parameters, including the declination, inclination, horizontal intensity, vertical intensity, total intensity, and north component X and east component Y along the horizontal direction. The three different components of MMM signals at our laboratory position measured by our self-developed instrument are 26.22 A/m, −2.57 A/m and 30.05 A/m, respectively, and the total magnetic field intensity is 39.96 A/m. This indicates that the testing results measured by self-developed instrument are very close to the theoretical calculation based on the world magnetic model, which can characterize the geomagnetic field very well.

Table 14.2 Geomagnetic field parameters at the laboratory position

Parameters	Declination (+East/−West)	Inclination (+Down/−Up)	Horizontal intensity (A/m)	North component X (A/m)	East component Y (A/m)	Vertical intensity (A/m)	Total intensity (A/m)
Value	−5.5911°	48.7536°	26.34	26.22	−2.57	30.05	39.96
Uncertainty	0.31°	0.21°	0.101	0.104	0.075	0.125	0.115

14.5 Testing of the Self-developed Instrument

14.5.1 Testing Method and Process

To verify the performance of the self-developed MMM testing instrument, the magnetic signals of 45 carbon steel were collected under the effect of geomagnetic and applied magnetic fields. The length, width and height of the 45 steel sample were 200 mm, 30 mm and 6 mm, respectively. A rectangular groove with a width of 10 mm and a depth of 2 mm was prepared in the middle of the sample. The symmetry axis of the sample was selected as the detection line with a length of 150 mm. The self-developed MMM testing instrument is shown in Fig. 14.5a. The whole MMM detection system is constructed as shown in Fig. 14.5b, which mainly includes computer, MMM testing instrument, DC stabilized current source, sensor support and Helmholtz coil. After detection, the three different components of the magnetic signals were analyzed further.

The detection probe is fixed on the support rod of the three-axis motion platform. The movement speed of the probe is 5 mm/s. The detection step is set to 1 mm, and the lift-off value is 2 mm. The detection steps are as follows:

(1) The sample is demagnetized by a TC-5 demagnetizer.
(2) The sample is placed on the test platform, and the self-developed MMM testing instrument is used to collect the magnetic signal along the testing line under the geomagnetic field. The data collected by the instrument are displayed on the screen.
(3) To reduce the external interference during the experiment, the detection system needs to be used in the area without electromagnetic interference. Based on the testing method in Chap. 8, the sample is placed into the Helmholtz coil system horizontally, and the direction of the magnetic field generated by the coil is perpendicular to the geomagnetic direction.
(4) The sample is magnetized by a Helmholtz coil with a magnetic field intensity of 12 Gs, and the detection process is repeated as shown in step (2).

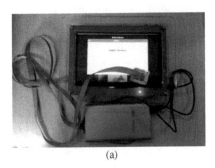

(a)   (b)

Fig. 14.5 Experimental instrument: **a** the self-developed MMM testing instrument; and **b** the whole detection system

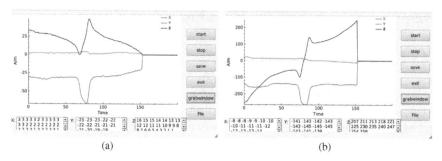

Fig. 14.6 MMM signals detected: **a** after demagnetization, and **b** after magnetization

14.5.2 Display and Analysis of MMM Signals

The magnetic signals of the sample collected by the self-developed MMM testing instrument were plotted in the display interface, as shown in Fig. 14.6, where X is the horizontal component of the magnetic field perpendicular to the detection line and parallel to the sample surface, Y is the magnetic field parallel to the detection line, and Z is the magnetic field perpendicular to the sample surface. Figure 14.6a and b show the detection results under the effects of demagnetization and applied magnetization, respectively. In Fig. 14.6a, the absolute value of the Y-direction magnetic signal at the rectangular groove defect reaches a maximum of 75 A/m. The Z-direction magnetic signal shows a peak-valley characteristic. In Fig. 14.6b, the absolute value of Y-direction magnetic signal at the defect also reaches a maximum of 300 A/m, which is higher than that in Fig. 14.6a. In addition, the Z-direction magnetic signal changes from negative to positive with a more obvious peak-valley characteristic. This indicates that the applied magnetic field can strengthen the characteristic amplitude of MMM signals at defects which has been discussed in Chap. 8. However, the X-direction magnetic signal always fluctuates at approximately 0 A/m, and the signal variation at the defect is not obvious. Therefore, the X-direction magnetic signal cannot be used to analyze the damage degree. It can be determined that the Y-direction magnetic signal is the tangential component $H_p(x)$ and the Z-direction magnetic signal is the normal component $H_p(y)$.

14.6 Comparison of the MMM Testing Instruments

To further verify the performance of the self-developed three-dimensional weak magnetic testing instrument, its specification is compared with the existing commercial MMM testing instruments (TSC-2M-8 of Russian Energodiagnostika Co. Ltd. and the EMS-2003 of Xiamen Edson Company), as shown in Table 14.3. The self-developed MMM testing instrument with a resolution of 0.4 A/m and sensitivity

Table 14.3 Comparison of the MMM testing instruments

Specification	Self-developed	EMS-2003	TSC-2M-8
Range of the magnetic field (A/m)	±644	–	±2000
Number of channels	16	8	8
Resolution (A/m)	0.4	–	1
Sensitivity (%FS/Gauss)	±2%	–	–
microprocessor (bit)	16	16	16
Memory capacity (Mb)	512	16	1
Flash memory capacity (GB)	4	2	–
LCD (lattice)	1366 × 768	320 × 256	320 × 240

of ±2% is more capable of sensing weak changes in magnetic signals in ferro-magnetic materials than the other instruments. The normal component and tangential component of the magnetic signals of the sample are also measured by the self-developed instrument and TSC-2M-8 commercial instrument, respectively, as shown in Fig. 14.7. The MMM signal characteristics obtained by these two instrument types are similar to each other. There is only a slight difference in the gain

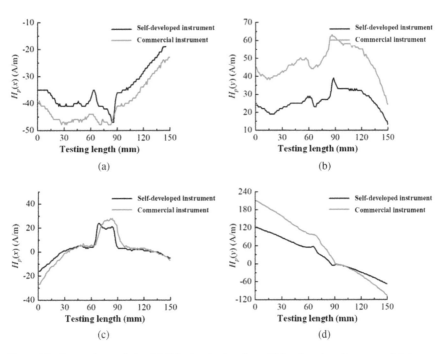

Fig. 14.7 Comparison of the MMM signals for the self-developed and commercial instrument: **a** tangential component after demagnetization, **b** normal component after demagnetization, **c** tangential component after loading, and **d** normal component after loading

coefficient during these two testing processes. This may be because the TSC-2M-8 commercial instrument cannot simultaneously detect all three-dimensional magnetic fields, which neglects the horizontal component of the magnetic field perpendicular to the detection line and parallel to the sample surface. Besides, it can be found that the overall amplitude of MMM signal remains at the level of geomagnetic field and fluctuates slightly after demagnetization. However, when the sample is subjected to applied loads, the sharp variations of MMM signals appear at the damage zone where the gradient of normal component increases and the peak characteristic of tangential component occurs. It indicates that the self-developed instrument can well determine the damage zone in ferromagnetic materials.

14.7 Conclusions

In this chapter, a new MMM testing instrument is developed based on the HMC5883 magnetic sensor and embedded system, and a graphical display interface is designed by Qt development tool to display the magnetic signal variation. The advantages of the self-developed MMM testing instrument are analyzed by comparing it to existing commercial MMM testing instruments. The results show that the self-developed MMM testing instrument can simultaneously measure all three-dimensional MMM signals in space with high accuracy, which can be widely applied to evaluate the damage degree of remanufacturing cores and the repair quality of remanufactured components.

References

1. A. Dubov, A. Dubov, S. Kolokolnikov, Application of the metal magnetic memory method for detection of defects at the initial stage of their development for prevention of failures of power engineering welded steel structures and steam turbine parts. Weld. World **58**, 225–236 (2014)
2. T. Werner, K. Rolf, S. Manfred, Process-integrated nondestructive testing of ground and case hardened parts. Eur. Federation Non-Destruct. Test. **8**, 1–5 (2002)
3. W.M. Zhang, C.F. Chen, H.Y. Li et al., Investigation of array magnetic memory testing technology on extension of metal crack. Appl. Mech. Mater. **148–149**, 1483–1486 (2011)
4. L.M. Li, B. Hu, S.L. Huang et al., A palm metal magnetic memory testing instrument. Nondestruct. Test. **26**(5), 249–252 (2004)
5. J.L. Ren, J. Wang, New method for metal magnetic memory quantitative analysis. Chin. J. Sci. Instrum. **31**(2), 431–436 (2010)

Lightning Source UK Ltd.
Milton Keynes UK
UKHW020736170522
403097UK00002B/9